Praise for [partially obscured by sticker]

Global Weirdness

"Written in straightforward prose and fact-checked by the world's eminent climate scholars, *Global Weirdness* reads like the 9/11 Commission Report: all of the facts, none of the hyperbole. In four succinct sections, its authors detail the truth about climate change."

—CBS SmartPlanet

"This primer on the science of global warming provides a fact-filled explanation of how climate change impacts, and will continue to impact, our daily lives. The sixty concise and easily digestible chapters tackle such questions as: Is climate ever 'normal'? What risks does climate change pose for human health? What are the economic costs and benefits of reducing carbon emissions? The authors are up-front about the potential downfalls of alternative energy and technological fixes."

—*Conservation Magazine*

"Without talking down to readers, the authors do a masterful job of clarifying all aspects of a complicated and alarming topic, making it that much more difficult for global-warming denialists to keep their heads in the sand."

—*Booklist* (starred review)

"An ideal introduction to the facts about global warming. . . . Lucidly written and thoughtful."

—*Kirkus Reviews*

"With quippy titles, helpful summaries, and a jargon-free writing style, Climate Central integrates scientific, historical, and sociological facts in an appealing and informative manner. . . . A great starter text on climate-change issues—fans of Bill McKibben will enjoy this work and then pass it along to skeptical friends."

—*Library Journal*

"An easily digestible read, with most chapters less than three pages long. Divided into four sections ('What the Science Says,' 'What's Actually Happening,' 'What's Likely to Happen in the Future,' and 'Can We Avoid the Risks of Climate Change?'), the book covers all the basics, including descriptions of Earth's previous climates and how hard it is for different cultures to adjust to changes; the difference between weather and climate; the greenhouse effect; and how climate scientists' predictions are coming true."

—*Publishers Weekly*

CLIMATE CENTRAL

Global Weirdness

This book was produced collectively by scientists and journalists at Climate Central, a nonprofit, nonpartisan science and journalism organization. The book was written by Emily Elert and Michael D. Lemonick; prior to external scientific peer review, it was reviewed by staff scientists Philip Duffy, Ph.D. (chief scientist); Nicole Heller, Ph.D. (ecosystems and adaptation); Alyson Kenward, Ph.D. (chemistry); Eric Larson, Ph.D. (energy systems); and Claudia Tebaldi, Ph.D. (climate statistics).

www.climatecentral.org

Global Weirdness

Severe Storms,
Deadly Heat Waves,
Relentless Drought,
Rising Seas,
and the Weather of the Future

CLIMATE CENTRAL

VINTAGE BOOKS
A Division of Random House, Inc.
New York

FIRST VINTAGE BOOKS EDITION, MAY 2013

The Library of Congress has cataloged the Pantheon edition as follows:
Global weirdness : severe storms, deadly heat waves, relentless drought, rising seas, and the weather of the future / Climate Central.
p. cm.
Includes bibliographical references.
1. Climatic changes. 2. Climatic changes—Forecasting.
3. Climatic changes—Mathematical models.
4. Greenhouse gases—Environmental aspects.
5. Global environmental change. 6. Global warming.
7. Weather forecasting. I. Climate Central, Inc.
QC903.G58 2012 577.2'2—dc23
2011047699

Vintage ISBN: 978-0-307-74336-7

Book design by Soonyoung Kwon

www.vintagebooks.com

Printed in the United States of America
10 9 8 7 6 5 4 3 2 1

Contents

Global Weirdness

Introduction

In February 2010, Thomas Friedman made the following plea in his *New York Times* column:

> Although there remains a mountain of research from multiple institutions about the reality of climate change, the public has grown uneasy. What's real? In my view, the climate-science community should convene its top experts—from places like NASA, America's national laboratories, the Massachusetts Institute of Technology, Stanford, the California Institute of Technology and the U.K. Met Office Hadley Centre—and produce a simple 50-page report. They could call it "What We Know," summarizing everything we already know about climate change in language that a sixth grader could understand, with unimpeachable peer-reviewed footnotes.

We couldn't agree more. It's quite remarkable that despite the steady growth in scientific understanding about the causes and effects of climate change, and the growing confidence of climate scientists that it poses a potentially serious threat to people, property, and ecosystems, the public seems more confused than ever. Is climate change really happening? If so, and if it's happened due to natural causes in the past, why should we think it's our fault this time? Haven't scientists been wrong before? They can't even predict the weather a week in advance; how can they possibly say anything about what the climate will be like fifty years from now?

A big part of the problem is that climatology is a relatively young and evolving field. Scientists are still learning about Earth's climate system—about how the land, oceans, and atmosphere absorb heat from the sun and move that heat around, and about how heat drives storms, droughts, sea-level rise, heat waves, and more.

But just because they don't know everything about the climate doesn't mean they know nothing. Far from it. They know for certain (and they've known for more than a hundred years) that carbon dioxide (CO_2) in the atmosphere traps the sun's heat. They know that burning fossil fuels including coal, oil, and natural gas adds extra CO_2 to the atmosphere beyond what's already there naturally. They know that humans have been burning more and more fossil fuels since the Industrial Revolution and that, as a result, levels of CO_2 in the atmosphere are more than a third higher than they were a couple hundred years ago. No responsible scientist, including most of those who have been labeled "climate skeptics," argues with any of this.

There's also very little argument over what the broad effects of an increase in CO_2 should be. The planet should get warmer. Sea level should begin to rise as warming ocean waters expand and as the warmer air melts glaciers and ice caps. That is exactly what both ground-based and satellite measurements have shown. On average, the oceans are about eight inches higher than they were in 1900, and the temperature is about 1.4°F hotter.

Things get more complicated when scientists try to predict what's likely to happen in the future. The reason is that Earth doesn't just respond passively to increasing temperatures: it can react in all sorts of ways that might boost the temperature rise or hold it back—and scientists haven't yet unraveled all of these possibilities. Increasing cloud cover could reflect extra sunlight back into space. Decreasing ice cover in the Arctic could do thc opposite. Melting Arctic permafrost might release extra carbon that has been in a deep freeze for hundreds of thousands of years. It's also not clear precisely how the changes in temperature will translate into changes in local conditions, although it's very likely that familiar weather and climate patterns will change, perhaps in surprising ways. That's why this book isn't titled "Global Warming," but rather "Global Weirdness," since warming is only part of what we can expect.

These uncertainties are one reason the Intergovernmental Panel on Climate Change, or IPCC, could only narrow the likely temperature rise by 2100 to between 3.2°F and 7.2°F above what it was in 2000. Another reason is that we don't know if fossil-fuel use will keep going up, or level off, or decline over that period.

This isn't to say that literally every climate scientist agrees with these findings. Some think that the temperature rise will be less than 3.2°F, while others think it could be more than 7.2°F. But there's no field in science, from genetics to evolutionary biology to astrophysics, where agreement is absolute. The reports issued periodically by the IPCC are meant to be snapshots of what climate scientists *generally* agree on at a given time (the most recent report came out in 2007; the next one is due out in 2013 or 2014). And despite some very public criticisms about the organization and its procedures, several independent investigations have shown only a tiny handful of scientific errors in the thousands of pages that make up the reports themselves. The same is true of the so-called Climategate episode, in which a few scientists said intemperate things in private e-mails and were somewhat sloppy in their record keeping. Outside investigators have found them guilty of carelessness but didn't find anything to cast doubt on the science itself.

Responsible scientists also know that it's important to keep questioning their own results. "The first principle," the physicist Richard Feynman once said, "is that you must not fool yourself—and you are the easiest person to fool." He meant that scientists need to consider all plausible explanations for what they observe, not just the most obvious or conventional. If Earth is warming, it's probably due to greenhouse gases, but it could instead be that the sun is putting out more heat. Scientists have looked carefully at that possibility, and it doesn't seem to hold up. They've also looked at the role of volcanoes and other natural factors that have caused warming or cooling in the past, and so far nothing explains the warming as well as greenhouse gases do.

Finally, the public has undoubtedly been confused by statements about climate change that sound authoritative but are simply false. Take the often-repeated assertion that global warming stopped in 1998. If you look at a graph spanning the years 1998–2010, that might appear to be close to the truth. But 1998 was an unusually warm year, so it's a misleading starting point. If you start in 1997 or 1999, things look very different. And if you zoom out to look at a graph spanning the years 1900–2010, it's clear that the first decade of the twenty-first century is warmer than any decade during that 110-year period.

All of this wouldn't matter very much if we were talking about a field like astrophysics. It ultimately doesn't matter whether there's a black hole in the center of the Milky Way or not. But if the effects of climate change are going to be truly disruptive, the problem would be dangerous to ignore. If they're not, we risk diverting a lot of resources for no reason. The difficulty is that if we wait until scientists are absolutely certain about every detail, it will be impossible to undo the damage, whatever it turns out to be.

So it's crucial for the public and for policy makers to understand what we do know about climate change; what we strongly suspect to be true, based on the available evidence; and what we're still uncertain about. Such knowledge is necessary to make informed decisions.

This book is an attempt to do just that: to lay out the current state of knowledge about climate change, with explanations of the underlying science given in clear and simple language. It's not exhaustive, but it covers the essentials. Since many aspects of the climate system are interconnected, so are many of the chapters: some of the information in the book appears in some form in more than one chapter.

In order to be as credible as possible, we've taken great care to avoid bias. We acknowledge that some aspects of the problem can't yet be addressed with certainty. We also make clear what climate scientists *do* know with a high degree of confidence.

To ensure technical accuracy, each chapter has been carefully reviewed internally by Climate Central scientists and revised in response to their comments. The chapters have then been reviewed again by eminent outside scientists who have particular expertise in the relevant subject areas—and then, if necessary, revised again.

The result, we believe, is an accurate overview of the state of climate science as it exists today.

A final note: we can't promise that all sixth graders will understand every word of this book. But we've tried to keep the language as simple, straightforward, and jargon-free as possible. We hope you find it useful.

WHO WE ARE

This book was produced collectively by scientists and journalists at Climate Central, a nonprofit, nonpartisan science and journalism organization. The book was written by Emily Elert and Michael D. Lemonick; prior to external scientific peer review, it was reviewed by the staff scientists Philip Duffy, Ph.D. (chief scientist), Nicole Heller, Ph.D. (ecosystems and adaptation), Alyson Kenward, Ph.D. (chemistry), Eric Larson, Ph.D. (energy systems), and Claudia Tebaldi, Ph.D. (climate statistics). For a list of outside scientific referees, please see page 213.

I

WHAT THE SCIENCE SAYS

1

"Normal Climate" Meant Something Different to the Dinosaurs and the Woolly Mammoths than It Does to Us.

Somewhere around 650 million years ago, geologists believe, long before the dinosaurs first appeared, glaciers covered Earth. This was far more intense than the much more recent ice ages (technically known as glacial periods), when Neanderthals and woolly mammoths thrived and glaciers crept down to smother large parts of North America and northern Europe. During this so-called Snowball Earth period, the land was buried under glaciers, from the poles right down to the tropics. It was so cold at the equator that the oceans may have been covered with a thin layer of ice (and maybe more) year-round.

Five hundred million years later, as the Cretaceous period was reaching its peak, there was practically no ice anywhere on Earth. Dinosaurs stomped around Antarctica, palm trees flourished in Siberia, and crocodiles slithered across Svalbard, north of the Arctic

Circle. During the Snowball, a significant fraction of the world's water was locked up in ice. During the Cretaceous, with all that ice gone, sea level was hundreds of feet higher than it is today.

Two million years ago, Earth entered yet another phase, known as the Pleistocene, during which the climate swung back and forth between glacial periods (what we think of as ice ages) and interglacials, when temperatures were roughly the same as they are now. The lengths of these periods varied, but they lasted some tens of thousands of years. We're now in an interglacial period, known as the Holocene, which started about ten thousand years ago.

These are only a few of the different climates Earth has experienced. Just before the glacial/interglacial cycles began two million years ago, for example, the Northern Hemisphere was free from glaciers, but Antarctica had lots of ice. During a bizarre episode called the Paleocene-Eocene Thermal Maximum, about fifty-five million years ago, the temperature went up by around 11°F, stayed high for a hundred thousand years, then dropped back again. If you look over the whole sweep of the planet's history, it's clear that "normal" climate has meant all sorts of different things at different times.

All of these changes have been "natural," since people didn't begin emitting greenhouse gases until about two hundred years ago. But of course the fact that climate can change naturally doesn't mean it can't also change as a result of human activity.

2

The Climate Has Changed Dramatically in the Past.

The heat-trapping properties of greenhouse gases are easy to measure in the laboratory, but those measurements don't always tell you what will happen in practice. Out in the world, a gas like CO_2 does trap heat. But the extra heat leads to other effects like melting snow and ice in the Arctic, more clouds, changes in vegetation, and many other things that could either increase or limit the amount of trapped heat.

To predict what might happen in the next hundred or even the next thousand years, scientists need to understand the total effect of extra greenhouse gases, not just the number you get in a lab. One way of doing that is to look deep into the past, to eras when it was warmer than it is now and periods when it was colder. If scientists can get an idea of how greenhouse gases (and other forces) affected temperature, sea level, and so on, long ago, they can get a sense of what might be coming.

There were no thermometers beyond a few hundred years ago, and nobody went out and took air samples. But there are plenty of natural records that can provide an indirect measurement of ancient conditions. They're known as proxies, and they're imperfect: they cover different periods of history and different parts of the planet, and some are more accurate than others. Even so, they give important clues to what happened to Earth's climate during times when greenhouse gases were more or less concentrated in the atmosphere than they are now.

Some of the most useful proxies come from the ice deep inside glaciers and ice caps, especially in Antarctica and Greenland. Bubbles frozen in the ice preserve samples of air that were captured up to 800,000 years ago by layers of snowfall compressed into ice by each year's successive layer of snow. When scientists test those air bubbles, they get a tiny whiff of greenhouse-gas concentrations from thousands of centuries ago. They also find dust particles and bits of plant material that offer clues about how wet or dry the land was and what sort of vegetation thrived at the time. Finally, the ice itself has a different composition when the global temperature is warmer or cooler: it's all frozen H_2O, but the O comes in slightly different varieties (a lighter and a heavier atom), and there's more of the heavier variety when air temperatures are higher.

Another set of proxies lies under the bottom of the sea. Things are constantly sinking to the seafloor, including dust and plant matter but also the remains of tiny shelled organisms known as foraminifera, or forams for short. Their shells are made of calcium carbonate, a mineral composed of oxygen, carbon, and calcium. The oxygen, just like the oxygen in ancient ice, comes in different weights depending on how warm the sea was when the shells were formed. The carbon also comes in different weights, and the mix is a clue to how much CO_2 was in the atmosphere. Whereas ice records go back less than a million years, the foram record reaches back tens of millions.

That's just a small sampling of the proxies scientists use to reconstruct past temperatures. Another is tree rings, which mark each year's growth throughout a tree's life but which can be thicker or thinner, depending on temperature and precipitation in a given year. Also corals, which build their shells from calcium carbonate; markings on rocks and on cave walls that show where sea level was at different times in the past (high sea level means it was warmer); journals and other written accounts from people who noted weather events centuries ago; and many more.

Again, all of these proxies are imperfect. When they're combined, however, they illustrate a generally consistent picture showing that when CO_2 levels are high, the global temperature is also high. They give us an idea of how the planet's temperature, atmosphere, precipitation, and other things behaved before we could measure them directly.

3

Our Ancestors Survived Climate Change. But It Wasn't Always Pretty.

The human species has lived through times of drastic climate change. Since modern humans first evolved, perhaps 200,000 years ago, the planet has gone through at least one Ice Age, when temperatures were somewhere between 7.2°F and 14.4°F colder than they are today (this is a worldwide average; some places would have been considerably warmer or cooler). That didn't stop our ancestors from spreading over most of the world. It didn't stop them from perfecting crude stone tools. It didn't stop them from creating beautiful paintings, like the ones on cave walls at Lascaux, France, or on rocks in Australia.

And when the climate warmed up dramatically, about ten thousand years ago, our ancestors not only survived but started to flourish. It's probably no coincidence that human civilization developed during these last ten thousand years, a period in which temperatures have been not only warm but extraordinarily steady. The climate was unusually stable during this period, at least in some parts of the world, which may have helped spur the development of agriculture. That in turn allowed for a dramatic expansion of our population.

Indeed, the human population has exploded over the past ten thousand years, from about five million at the start (about the same as the Washington, D.C., metropolitan area, but spread across the world) to nearly seven billion today.

Beyond that, the original five million had no permanent addresses. Just as their ancestors had for millions of years, they lived in small bands that might stay in one place for an extended time, but if they had to pick up and move because a glacier was inching closer or a drought was lingering too long, they could do it. Some of these bands were undoubtedly wiped out by changes in climate, among other reasons. However, these events wouldn't threaten the species as a whole.

But after our ancestors invented agriculture, most people stopped foraging through the fields and forests for fruits and seeds and began planting them. Farming is a much more efficient way of getting food, and an abundant food supply made it possible for populations to grow. However, it also meant people had to stay in one place, which led to the creation and growth of cities. These tended to arise near lakes, rivers, and oceans—sources of food, water, and transportation—which are also likely to rise or fall as the climate changes.

That's why climate change is such a worry today. It's one thing for a small band of people to pack up camp and move a couple hundred miles to a better location if the climate changes. It's a very different thing to try to move a city like Cairo or New York or Shanghai because the sea level is rising. It's very different to relocate the farms of the Midwestern United States up to Canada—along with the highways and railroads and power lines that serve them—because it's become too hot and dry to grow grain. It's very different to move tens of millions of people out of harm's way in low-lying coastal nations like Bangladesh. Our civilization, in short, is very highly adapted to the present climate. Major changes of any sort would almost certainly be extremely disruptive.

So while the human species has survived climate change in the distant past, those adjustments were relatively simple—but even so, they probably involved levels of mortality that we are unaccustomed to today. It should be a given that individual lives matter, not just the survival of the species. The result of global warming is projected to push temperatures above the range of previous interglacial periods, but even if the changes were no greater than those humans have lived through in the past, our very large population and the existence of modern civilization make this a whole different ball game.

4

Dinosaurs Didn't Drive Gas-Guzzlers or Use Air-Conditioning.

Earth's climate has changed continually since the planet was formed, sometimes dramatically. As we've seen, scientists know this from, among other things, looking at air bubbles trapped in ice and at layers of ancient mud at the bottom of the ocean. Obviously, humans couldn't have caused these changes, since we've only been around for a tiny fraction of our planet's history.

Scientists have identified a number of these natural forces, which cause both long, gradual swings and short, sudden shifts in climate and include changes in the sun's energy, volcanoes, wobbles in Earth's orbit, and changes in ice sheets and ocean currents.

Billions of years ago, when the planet was young, the sun was putting out only about 70 percent as much energy as it is now. This should have frozen all the water on Earth, yet we know that there was liquid water back then, so it couldn't have been that cold. Scientists believe the answer to what's known as the "faint young sun paradox" is that there was a lot more heat-trapping CO_2 in the atmosphere at the time. As the sun gradually got brighter over billions of years, CO_2 levels dropped in a way that kept temperatures relatively fixed. This natural thermostat won't have any effect on the current episode of global warming, however, because it operates very, very slowly.

Volcanic eruptions can also cause warming because they spew extra CO_2 into the atmosphere. They're not the cause of recent climate change, since volcanoes now emit less than 1 percent as much CO_2 as humans do. But CO_2 stays in the atmosphere for a long time, so CO_2 from volcanoes building up over millions of years can be an important factor controlling Earth's climate.

Volcanoes may also cause brief cooling. Some very powerful eruptions throw sulfur dioxide into the upper atmosphere, where it can spread rapidly, reflect incoming sunlight, and cool Earth. The eruption of Mount Pinatubo in 1991, for example, cooled the planet by about 1°F for the next couple of years. In 1992, the United States experienced its coldest, wettest August in seventy-seven years. The sulfur dioxide lasts only a couple years, though, after which the planet warms up again. (Truly gigantic eruptions millions of years ago caused much more dramatic cooling.)

Earth's motion can also trigger climate change. The shape of Earth's orbit around the sun varies slightly, becoming more oval and then more circular again, over about a hundred thousand years. The planet also wobbles like a top, and rocks slightly back and forth, in cycles lasting thousands of years. All of these change the amount of solar energy reaching different parts of Earth at different times of year.

These variations are not enough to cause dramatic climate change on their own. But they do trigger natural processes that amplify their effects by means of what's known as a feedback. One of these involves ice. Because of their bright white color, ice sheets help keep the planet cold by reflecting sunlight back into space. When the orbital changes cool Earth a little bit, extra ice forms, which reflects more light and cools the planet further, leading to more ice, and so on. When the orbital changes warm the planet, some of the ice melts. Less energy gets reflected, which lets the planet warm up even more, melting more ice.

Another feedback is CO_2. When the atmosphere gets colder, the oceans absorb extra CO_2, leaving less in the atmosphere to trap heat. That makes it even colder, which lets the oceans absorb even more. When it gets warmer, the process goes in reverse, just as it does with ice feedbacks. So CO_2, like ice sheets, is both a cause and an effect of temperature changes.

The amount of carbon dioxide in the atmosphere goes up and down naturally for other reasons, too. Plants, for instance, absorb carbon dioxide from the atmosphere in order to grow. Although it may seem surprising, some rocks absorb CO_2 as well, although much more slowly than plants do.

Finally, shifting ocean currents can cause both regional and worldwide changes in climate. On a global scale, ocean currents act like a giant conveyor

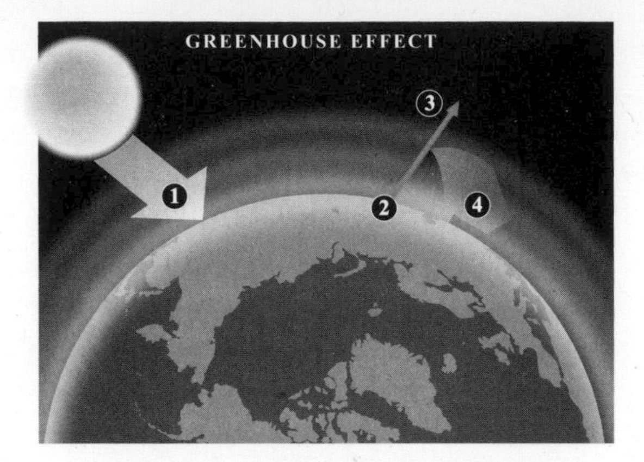

Sunlight passes freely through the atmosphere to warm the surface of the Earth (1). The Earth sends some of that warmth back outward in the form of infrared radiation (2), and some of it escapes into space (3). As we add greenhouse gases like carbon dioxide to the atmosphere, though, more heat is trapped (4), and the temperature rises.

belt, spreading heat around Earth. When this conveyor belt slows, there are major changes in regional climate.

The frequent ocean phenomenon called El Niño, which is best known for affecting regional patterns of temperature and precipitation, also changes global average temperatures and is caused by migrating bodies of warm seawater.

Scientists have no doubt, in short, that natural forces cause climate change. However, the fact that something can happen naturally doesn't mean it's always natural. Here's one way to think about it: For hundreds of millions of years, forest fires were triggered by lightning strikes, with no human involvement at all. Now, as arson investigators will attest, that's no longer the case for many forest fires.

5

Carbon Dioxide Is Like a Planetwide Sweat Suit (Sort Of).

The atmosphere's heat-trapping process is known informally as the greenhouse effect, because heat is trapped by molecules in the atmosphere just as a greenhouse traps heat in its interior. Clearly what keeps the heat from escaping a greenhouse isn't a molecule: it's the glass walls and ceiling. But although the terminology may be imprecise, our understanding of this science is not, and claims of its existence aren't even slightly controversial. The British scientist John Tyndall proved it in a series of lab experiments in 1859—more than a century and a half ago.

Here's how it works. The sun bombards Earth with energy, including the light we can see but also other kinds (such as ultraviolet light) that we can't. Unlike the other forms, visible light passes easily through Earth's atmosphere to heat up the land and the oceans. As the land and the oceans warm up, they release some of that heat. Part of the heat comes out as infrared radiation (another kind of light that's invisible to human eyes).

If Earth had no atmosphere (or a very thin one, like Mars), that heat would escape directly into space. But gas molecules in Earth's atmosphere keep that from happening. Carbon dioxide, or CO_2, is one of them. When infrared light hits a molecule of CO_2, the molecule absorbs the light. It's just a side effect of how the molecule is put together.

When it absorbs infrared light, a molecule of CO_2 heats up a little and then sends some of that heat back out. In a way, it's like a miniature version of what Earth as a whole does: heat up, then radiate some of that heat away. Because some of Earth's heat is caught by CO_2 molecules, it doesn't immediately escape to outer space: it first warms up other air molecules and ultimately warms Earth itself. (The same thing applies to other greenhouse gases, including methane, nitrous oxide, and water vapor.)

Without any heat-trapping CO_2 in the atmosphere, Earth's average temperature would be frigid—roughly 0°F. On Venus, where the thick atmosphere is almost all CO_2, so much heat gets trapped that the temperature hovers at a horrifying 890°F. On Mars, which has a very thin atmosphere, the average surface temperature is about -80°F. (Venus is closer to the sun than Earth, and Mars is farther away, which explains some of the difference, but nowhere near all of it.) The existence of a natural greenhouse effect on Earth and other planets is one reason why scientists are so confident that adding more greenhouse gases to Earth's atmosphere is heating things up.

6

"Global Warming" or "Climate Change"? Doesn't Matter, It's All the Same.

When CO_2 builds up in Earth's atmosphere, whether it's from natural causes or human activity, the greenhouse effect makes the temperature start rising. In other words, the globe warms—which is where the term "global warming" comes from. (Other natural causes can trigger global warming—and for that matter global cooling—but here we're talking about the effect of CO_2.)

But the result of a rise in temperature isn't just a slightly warmer version of the world we have now. Mountain glaciers will start to melt, so the rivers they feed will flow differently. More water will evaporate from both land and oceans, leading to more rain and snow in some places and more drought in others. As the oceans warm, major currents may speed up or slow down—which would affect weather patterns.

The temperature differences between the equator and the poles may change as well, because greenhouse warming isn't likely to be spread evenly across Earth.

Since differences in temperature are what makes the wind blow, the direction and force of the wind could change, thus altering weather patterns. Animals and plants that are comfortable in a particular geographic location might have to move somewhere else to find the conditions they are adapted to, or they might become extinct.

Those are just a few of the changes a warmer world is likely to experience. Beyond that, a buildup of CO_2 in the atmosphere would lead to another effect besides warming. Some of that extra carbon dioxide would be absorbed by the oceans, turning the water slightly more acid than it is now. That could also have a major effect on plants and animals that live in the seas.

Since many of the disruptions that could come from adding greenhouse gases go beyond just warmer temperatures, many scientists prefer to talk about climate change rather than global warming. That doesn't mean "global warming" is incorrect. It's just that "climate change" gives a more complete idea of what's going on.

Even then, there's a chance of confusion, because the climate has changed many times over Earth's history, for purely natural reasons. So to be technically correct, you have to distinguish between natural climate change and human-caused climate change (or "anthropogenic" climate change, from the ancient Greek words for "man" and "caused").

Nobody but a scientist is going to use the term "anthropogenic climate change," though. For most people, either "global warming" or "climate change" is about as technical as you need to get, and either term is fine. In common use, they both mean essentially the same thing.

7

Weather Is Not Climate. Climate Is Not Weather. Except They Kind of Are.

Everyone instinctively knows the difference between weather and climate. Both involve things like temperature, rain, snow, clouds, and wind. But weather is short-term: it applies to what's going on day to day, or even hour to hour. Climate is long-term: it describes the average conditions in a particular region over many years. That's why a single weather event can never prove or disprove the reality of human-caused climate change.

Weather can change very quickly as weather systems sweep across the country. It might be sunny and warm one day, then cool with torrential rains the next. Climate can't change like that. January in northern Vermont is about 52°F colder, on average, than July. If you want a place that's warm year-round, you don't go to Vermont. Vermont has cold winters: that's how it was last year, and you can count on it being that way next year, even though the temperature might change by a few degrees up or down from one year to the next.

But while Vermont winters are always a lot colder than Vermont summers, the day-to-day weather can bounce around quite a bit. The climate in Minneapolis, Minnesota, is frigid in winter and warm in summer. That doesn't mean we will never get a warm day in winter. The thermometer hit a record 64° one day in February 1896 and plunged to 39° in August 1967. Washington, D.C., doesn't usually get bad snowstorms, but in the winter of 2010 it got several of them. All of that is weather, not climate.

Climate can change too, but much more slowly. The average temperature or precipitation in any location might creep up or down over a period of decades, but the weather is always going to change more day to day than the climate changes year to year. A 100° day in New York in July is hotter than average, but by itself it doesn't mean the climate is getting warmer. A string of 100° days in New York in July doesn't necessarily mean the climate is getting warmer.

But if New York keeps getting more record-high temperatures decade after decade and fewer record-low temperatures, that's a hint that the climate might really be changing.

8

On Venus, the Greenhouse Effect Makes It Hot Enough to Melt Lead.

Venus is a bone-dry world where the ground is almost unimaginably hot—nearly 900°F, according to space probes that have taken the planet's temperature. If you dropped a lead brick onto the surface of Venus, it would melt into a puddle.

It's not surprising that Venus should be warmer than Earth. It's closer to the sun (about 30 percent closer, in round numbers), so it gets more heat. But that's not nearly enough to explain the temperature difference. Instead, planetary scientists think Venus is the victim of what's known as a runaway greenhouse effect.

What they think happened was that billions of years ago, Venus probably had plenty of liquid water on its surface, just as Earth does today. Since it's closer to the sun, the water on Venus evaporated easily, and since water vapor is a heat-trapping greenhouse gas, the atmosphere warmed up. That made for even more evaporation, leading to an even warmer atmosphere, and so on.

It gets worse. Venus's early atmosphere almost certainly had plenty of CO_2 in it as well. As long as there was water on the surface, the water would have absorbed some of that CO_2. The CO_2 would then have combined with minerals over millions of years to form rock. The same thing has happened to much of the CO_2 in Earth's original atmosphere.

Once the water was gone from Venus's surface, that couldn't happen anymore. So any CO_2 still in the atmosphere would have stayed there, and any extra CO_2 belched out by volcanoes, say, would have stayed as well. Most of the water vapor is now gone from Venus's atmosphere; it leaked out into space long ago. But the CO_2 is still there—so much of it that it makes up about 95 percent of Venus's atmosphere. And that's what keeps the planet so hot.

One obvious question is whether the same thing could happen here on Earth, where we have plenty of water for evaporation. Scientists think it probably couldn't. As Earth warms, extra water vapor will enter the atmosphere, and it will act as a greenhouse gas. But while this will amplify the warming effects of CO_2, calculations suggest it won't lead to a runaway greenhouse effect.

9

Carbon Dioxide Is Only Part of the Problem.

Carbon dioxide is a heat-trapping greenhouse gas, but it's not the only one. It's not even the most common. That honor goes to water vapor, which is the third most abundant gas of any kind in the atmosphere, after nitrogen and oxygen. There's about a hundred times more water vapor in the atmosphere, on average, than there is CO_2.

Carbon dioxide gets a lot of attention because humans are adding it to the atmosphere. Water vapor, on the other hand, comes pretty much entirely from evaporation from oceans, lakes, rivers, and soil. The amount of water vapor in the atmosphere depends almost entirely on how warm Earth is. If the planet gets warmer, say from adding CO_2 to the atmosphere, more water evaporates into the atmosphere, and the atmosphere can hold on to more of it. The extra water vapor traps more heat, and it gets even warmer. So water vapor amplifies the warming from human CO_2 emissions; it's another feedback. And even though water vapor is the most important greenhouse gas, we can't control how much of it there is in the atmosphere, except indirectly, by controlling the temperature.

The third most common greenhouse gas, after water vapor and CO_2, is methane. Like CO_2, some of it comes from natural sources—mostly bacteria, which decompose plant matter in swamps and help termites digest their food. Methane also comes from human activity. Oil and natural-gas drilling can release methane from underground. Landfills release methane from rotting garbage. Cattle burp out methane the same way termites do. Rice is grown in artificial swamps, which churn out methane just as natural swamps do.

After methane comes nitrous oxide, and just like CO_2 and methane it has both natural and human sources. The natural sources are, once again, bacteria, mostly in soil and in the ocean. The human sources: agriculture, from soil treated with nitrogen-rich fertilizer and churned up by plows. It also comes from car and truck exhaust, from making nitric acid in chemical plants, and from several other sources.

The list goes on, with lots of industrial chemicals (many of them containing the elements chlorine and fluorine) also acting as greenhouse gases.

It's important to realize that many of these gases are actually much more efficient, molecule for molecule, at trapping heat than CO_2—literally tens of thousands of times more efficient for some of the industrial chemicals. They'd be front-page headlines, but there's

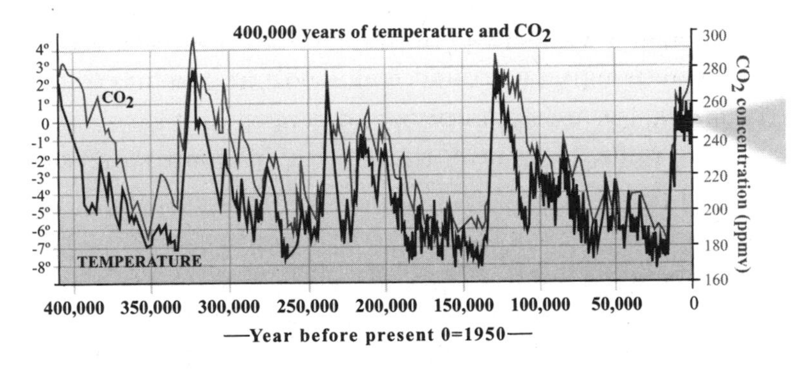

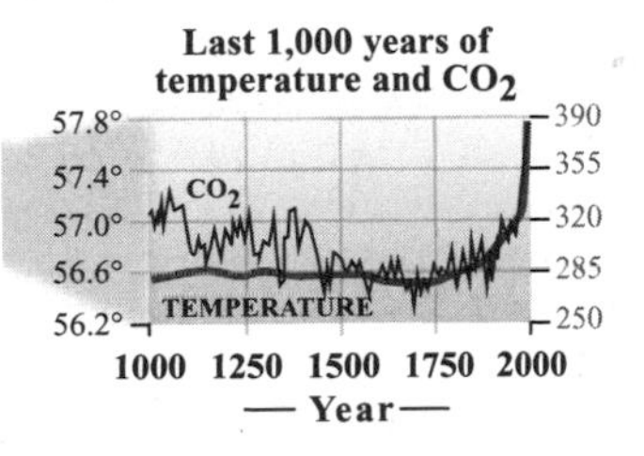

so much less of them in the atmosphere that their impact is relatively small. Methane is far more common than these chemicals, and it's twenty-one times as efficient as CO_2 at trapping heat, but it stays in the atmosphere for a relatively short time before breaking down. It's also important to realize that most of these other gases occur naturally. Their warming effect is part of why the planet is habitable. But just as with CO_2, pushing them beyond their natural levels can change Earth's energy balance.

So when climate scientists and policy makers try to figure out which greenhouse gases to be most concerned about, they have to think about how much of a particular gas there is, how much control humans have over emissions, how powerful the gas is, and how long it lingers in the atmosphere.

When you put all that together, CO_2 comes out first, by a lot, and that's why we hear most about it.

10

Once We Invented the Steam Engine, Climate Change Was Pretty Much Inevitable.

As we saw in chapter 2, scientists have many ways of figuring out how much carbon dioxide was in the atmosphere in the distant past—by looking at air bubbles trapped in ancient ice that preserve samples of the atmosphere from as long ago as 800,000 years or by examining forams at the bottom of the sea. More recently, around the late eighteenth century, people started taking direct air samples and measuring their CO_2 content. And of course, scientists do that today as well, with even greater accuracy.

Based on all of these ways of measuring, scientists agree that CO_2 levels varied between 260 and 280 parts per million, or ppm (that is, 260–280 CO_2 molecules for every million air molecules), during the most recent ten thousand years. Levels rose gradually over this time, settling at 280 ppm at the start of the Industrial Revolution in the nineteenth century. Some people argue that this slow preindustrial rise was a result of human activities, primarily clearing of forests for agriculture.

But in the late eighteenth century, the Scottish inventor James Watt patented an efficient version of the steam engine, which helped launch modern industrial civilization. Most steam engines run on coal or oil, which releases CO_2 when burned. The machines of the Industrial Revolution fueled a dramatic rise in economic activity, which triggered a huge population increase, from about 700 million people in 1750 to almost 7 billion today.

Nearly all of those people, especially in developed countries like the United States, use energy for transportation, heating, and—something new since 1900 or so—electricity. While steam engines have become mostly a quaint curiosity, most of the energy we use still comes from the burning of carbon-rich coal, oil, and natural gas.

It's not even a little bit surprising, therefore, that CO_2 levels in the atmosphere have climbed from about 280 ppm in 1800, at the beginning of the Industrial Revolution, to about 390 ppm in 2011.

11

The Ozone Hole Is Not Global Warming. Global Warming Is Not the Ozone Hole.

It's pretty common for people to confuse the ozone hole and the greenhouse effect, but while they're both caused by human activity, they're almost completely unrelated.

The ozone hole works like this: every spring (the Southern Hemisphere spring, that is, which begins in September), the atmosphere's ozone layer thins out in the skies above Antarctica and nearby areas. This so-called ozone hole, which repairs itself every year only to open again the following spring, was first discovered in the mid-1980s. Within a couple years, scientists had figured out what was causing it: a category of chemicals called chlorofluorocarbons, or CFCs. CFCs don't exist in nature; they were invented, in the 1920s, and proved incredibly useful for powering refrigerators and air conditioners and for making spray cans work.

CFCs are absolutely harmless—until, that is, they

get up into the stratosphere. Up there, they break apart and destroy ozone molecules (ozone molecules are made up of three oxygen atoms). They don't create an actual "hole"—more like a region where the ozone becomes sparse. It's a problem because the atmosphere's ozone layer helps shield Earth's surface from the sun's ultraviolet rays. Even with a normal amount of ozone, enough ultraviolet gets through to give you a bad sunburn. If the ozone hole were to spread much further, skin cancers could become a lot more common, and plants could be harmed. Fortunately, an international treaty called the Montreal Protocol banned CFCs in 1987, so the ozone hole should eventually go away.

So what does all of this have to do with global warming? Very little. It's true that CFCs are greenhouse gases, but they're minor compared with carbon dioxide, methane, and the other big players.

Still, it's understandable that people confuse the two. Both are potentially dangerous, both involve the atmosphere, and both are caused by human activity.

12

The Northern Hemisphere Has Heated Up More in the Past Half Century Than in Any Similar Period Going Back Many Hundreds of Years.

Earth's global temperature record goes back to about 1850. Thermometers were around for a long time before that (Galileo invented one at the end of the sixteenth century), but they were only used to take local temperatures in a few places. Scientists couldn't calculate the worldwide average temperature until they could get thousands of measurements from around the planet.

Based on measurements since the mid-nineteenth century, we know that Earth has gotten about 1.4°F warmer over the last hundred years. We also know that over the past fifty years, the temperature rose about twice as fast as it did over the fifty years before that.

Scientists have taken that temperature record and combined it with proxy data—changes in tree rings, ice layers in Greenland and Antarctica, and other natural phenomena that change as the temperature changes. Using all of that information, they've been able to estimate average temperatures over the entire Northern Hemisphere going much further into the past. The graph of those average temperatures has become known as the hockey stick, because it shows a long, mostly flat line for several hundred years and then a sharp upward curve during the last hundred years that looks roughly like the blade at the end of a hockey stick.

After the first versions of the hockey-stick graph were widely published, some critics argued that the temperature reconstructions were inaccurate and that the claim of unprecedented recent warming was an exaggeration. So scientists went back to recheck the original methods and try new ones. They did reconstructions of past temperatures both with and without some of the proxies (some types of tree rings, for example) that had come under question.

In all cases, the shape of the graph didn't change much. It still looked like a hockey stick, with fairly stable temperatures over hundreds of years, followed by a sharp rise during the twentieth century. A panel of independent experts brought together by the National Academy of Sciences, meanwhile, declared that the original hockey-stick graph—the one that had been criticized—had been valid all along.

Today, the vast majority of scientists agree that for at least the last half century, average temperatures in the Northern Hemisphere have been higher than they were for any previous fifty-year period during the last thirteen hundred years. This doesn't mean

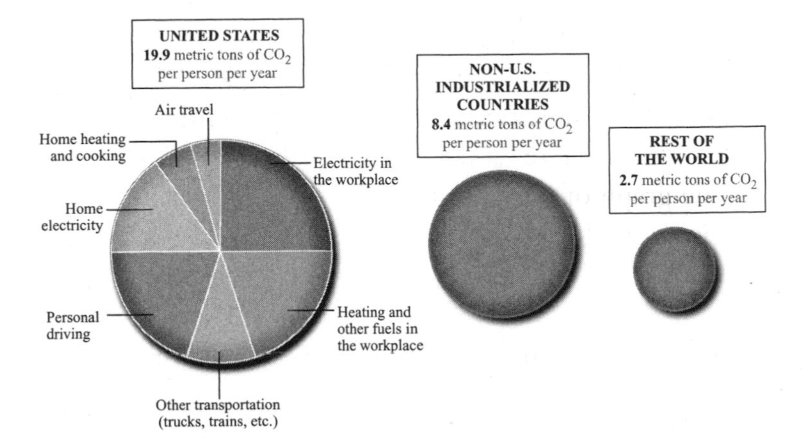

temperatures were perfectly steady all that time. The planet was probably warmer than average during the so-called Medieval Warm Period (around A.D. 1000) and cooler than average during the Little Ice Age (about A.D. 1500–1800).

But temperatures haven't risen as fast or gone as high over such a period as they have since about 1950.

13

Coal Alone Churns Out 20 Percent of Human Greenhouse Emissions.

Millions of years ago, large parts of the United States (and other parts of the world) were covered in dense, tropical forests. Then the oceans rose, and the forests were buried under the seafloor. Over time, the wood was transformed into the hard black carbon-rich substance we call coal. Then, at the end of the eighteenth century, the steam engine was invented, and before long we were burning coal for everything from powering railroads to generating electricity.

We're still burning it. Coal supplies the world with about 25 percent of its annual energy demand and the United States with half its electricity. The United States alone burns about one billion tons per year (that's about eighteen pounds, per person, every day). Coal is an attractive source of energy because it's relatively cheap and there's a lot of it. The United States is the second-largest producer of coal in the world after China, and at the rate we are using it now, we're still a couple hundred years away from digging up and burning through all the known coal in the country.

But there's a catch. Because coal is so rich in carbon, it generates a lot of carbon dioxide when it's burned—typically, about two pounds of CO_2 for every pound of coal. That's about twice the CO_2 that comes from generating an equal amount of energy from natural gas and about a quarter more than the equivalent amount of oil. Coal is the leading source of energy-related greenhouse-gas emissions, accounting for about 40 percent of the world's total. If current trends persist, that percentage is likely to grow in the future as energy demand increases.

14

A Quarter of the CO_2 in the Atmosphere Comes from Fossil Fuels, and It's on the Way Up.

Not all carbon dioxide is created equal, because the carbon atoms it contains come in slightly different weights. About 99 percent of the carbon in the atmosphere is carbon-12, the lightest kind. The rest is carbon-13 and carbon-14, which are a little heavier. Since humans started adding CO_2 to the atmosphere through fossil-fuel burning and deforestation, the percentage of these heavier carbon atoms in the atmosphere has gotten smaller, and that's a smoking gun that can tell us a lot about where the carbon came from.

Actually, it's two different kinds of smoking guns. What makes carbon-14 so informative is that it's radioactive (although there isn't enough of it to be at all dangerous). Over time it changes into nitrogen through radioactive decay. That means that the amount of carbon-14 stored in plant tissue gradually gets less over time (it takes nearly six thousand years for half the carbon-14 in a given sample to decay into nitrogen) and eventually becomes very small.

Fossil fuels such as coal and oil are mainly the decomposed remains of plants that lived millions of years ago. Because the remains are so old, just about all the carbon-14 has changed into nitrogen. So when we add CO_2 to the atmosphere by burning fossil fuels, the total amount of CO_2 increases, but the amount of carbon-14 does not. Since humans started burning fossil fuels over two hundred years ago, the fraction of carbon-14 in the atmosphere should be getting smaller, and that's exactly what air samples show. The exception was during the brief period when nuclear weapons were being tested in the atmosphere, creating more carbon-14. Measuring the shrinking percentage of carbon-14 helps confirm that fossil-fuel burning has been increasing the amount of atmospheric CO_2.*

As for carbon-13, it's not radioactive like carbon-14, but it's still informative. Carbon-13 molecules are bigger than carbon-12, so plants have a harder time turning them into food. For that reason, plants have a lower fraction of carbon-13 than the atmosphere does.

Because fossil fuels come from plants, they also have less carbon-13 than the atmosphere. So as we've added fossil-fuel carbon to the atmosphere, the fraction of carbon-13 should have decreased. And it has.

Tracking these two types of heavier carbon atoms tells us that about 25 percent of the CO_2 now in the

*Carbon-14 in the atmosphere decays the same way carbon-14 in fossil fuels does, but it's constantly replaced by cosmic rays striking the upper atmosphere from space.

atmosphere appears to have come from fuel burning or from the destruction of living plants, largely from the clearing of rain forests. That's about the same amount scientists estimate should be there based on how much fossil fuel we've burned. So they're even more confident that the extra CO_2 in the atmosphere comes from humans.

15

If We Stopped Burning Fossil Fuels, We'd Keep Emitting Greenhouse Gases.

As we've just discussed, humans have been adding a lot of extra carbon dioxide to the atmosphere since the invention of the steam engine, through the burning of fossil fuels. But while that accounts for a large fraction of the greenhouse gases we've generated during that time, it doesn't account for them all.

Much of the rest comes from deforestation, agriculture, and concrete manufacturing. Cutting down forests—done mostly to clear land for agriculture—releases CO_2 as plant matter decays or is burned. That process also releases large amounts of methane and nitrogen oxides, both of which are potent greenhouse gases.

Other agricultural practices also release greenhouse gases. Soils store a lot of carbon in the form of organic matter and release carbon dioxide as they're tilled and exposed to the air. The widespread use of commercial fertilizers creates significant emissions of nitrous oxide. Rice paddies generate large amounts of methane, and the digestive tracts of sheep, cows, and other livestock generate even more.

Another surprisingly big contributor to CO_2 emissions is the cement industry. It takes energy to make cement. Even more significant, cement is made from carbonate rocks, such as limestone, which as the name implies, contains carbon. When these rocks are heated as part of the cement-manufacturing process, they release carbon dioxide. All told, the cement industry alone accounts for about 5 percent of CO_2 emissions worldwide.

Finally, with a current world population of more than seven billion people, Earth has over six times as many humans living on it as there were in 1800. That adds up to a lot of waste: our landfills, some of which are big enough to be seen from space, are yet another hefty source of methane.

16

No Natural Force Has Been Able to Explain the Recent Warming.

Back to the planet's average temperature rise of about 1.4°F during the twentieth century. This long-term upward trend means one of three things: either Earth was receiving more energy from the sun, or Earth was receiving the same amount of energy but keeping more of it—or a combination of the two.

It turns out that the sun's energy output did in fact rise during the first half of the century. Then it stopped rising. And over the past few decades—the time when temperature rose the fastest—the sun's energy output actually decreased a bit. So that doesn't explain it.

Earth's temperature can also cool if some of the sun's energy is kept from reaching us. Volcanic eruptions can throw huge amounts of dust and gas into the atmosphere, including particles called sulfate aerosols. These aerosols form a haze in the stratosphere that keeps some of the sun's energy from getting through. The gigantic eruption of Indonesia's Mount Tambora

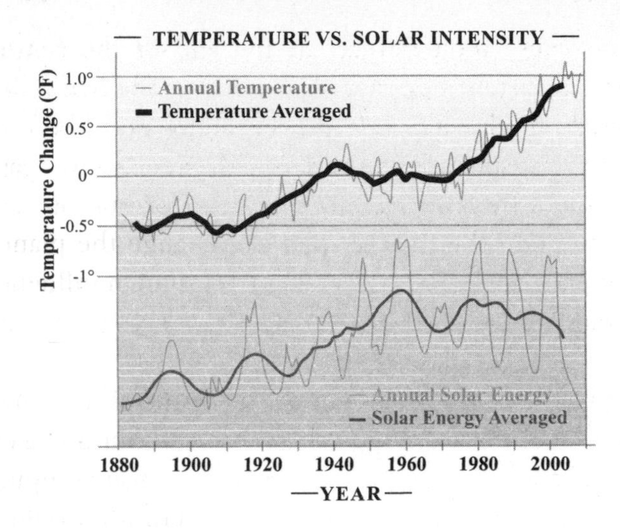

in 1815 blocked so much of the sun's energy for such a long time that 1816 became known as the "Year Without a Summer."

But if especially *few* volcanoes are erupting in a particular period, the reverse can happen: with fewer aerosols than normal to block the sun, more energy gets through than usual, and the temperature can rise.

Climate scientists have looked back over the twentieth century to see whether a lull in volcanic activity could be the main cause of the overall warming, but that doesn't explain it either. When they graph all of the known natural influences on global temperature, scientists would have expected the planet's thermometer to have risen in some decades and fallen in others. But roughly speaking, the global average temperature in any given twenty-year period, whether it was at the beginning, the middle, or the end of the century, should have been about the same as any other.

Because temperatures at the end of the century were consistently higher than those at the beginning, something else must have been going on. The most obvious suspect is heat-trapping greenhouse gases generated by humans, along with deforestation, pollution, and other factors that can change the planet's energy balance. We know that these human influences on climate grew at a faster and faster rate as the century went on.

But having a suspect isn't the same as making a case. To understand the effect of greenhouse gases and all the other factors, scientists can use computer models to calculate what Earth's temperature would be under different scenarios—including some that take into account human influences and others that leave them out. Then they can compare conditions in these virtual worlds with what actually happened in the real world.

It turns out that the only way to get temperatures in the models to match the long-term behavior of temperatures in the real world is to include human influences along with natural forces. Similar experiments have looked at other aspects of the climate, including extreme weather events and the fact that warming is happening at different rates in different places. In all cases, the only way to explain the observations is to include the effects of human activities—especially human-generated greenhouse gases. For that reason, most scientists agree that those forces are now the major factor controlling the warming trends—and the many other related changes—we've seen across the world over the past few decades.

17

CO_2 Could Stay in the Air for Hundreds or Thousands of Years, Trapping Heat the Whole Time.

Before the Industrial Revolution, natural processes were sending more than 700 billion tons of carbon dioxide into the atmosphere every year. Now the fossil fuels we burn are adding about 30 billion more. It adds up to only about 4 percent of the total, but those extra emissions make a big difference.

Before humans started adding CO_2 to the system, carbon dioxide that entered the atmosphere naturally was removed by natural processes at about the same rate: it was absorbed by the oceans and by plants, or it reacted chemically with rocks. This is why the level of CO_2 in the atmosphere has remained stable over time. But since people started burning tons of coal and other fossil fuels during the Industrial Revolution, those natural processes haven't been able to keep up. As a result, the amount of CO_2 in the atmosphere has been on the rise.

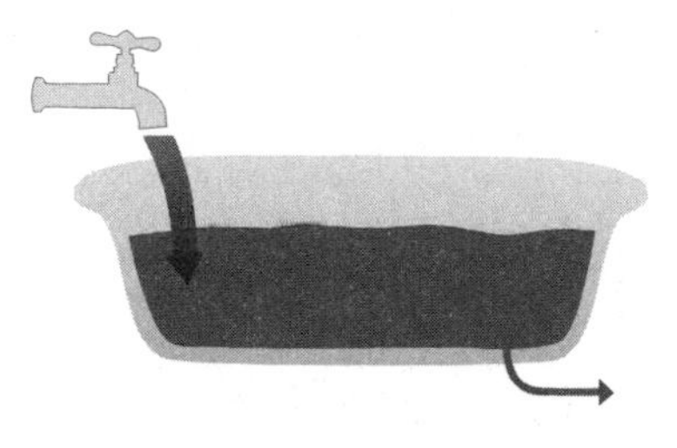

It's something like a bathtub with a slow drain. It might be gradual, but if the faucet is turned up just a little too high, you're going to get halfway through your shower and realize you're up to your ankles in water—and the water is still rising.

The same sort of thing happens with carbon dioxide. Humans have turned up the faucet by pumping extra CO_2 into the atmosphere. Eventually, if we stopped adding CO_2, the excess would be removed by plants, oceans, conversion into minerals, and other processes. But just as with the bathtub, the more the CO_2 builds up, the longer it takes to go away. (In fact, it's worse than the slow-drain problem, because the warmer Earth gets, the more slowly CO_2 is absorbed by oceans and plants.)

As a result of turning up the CO_2 faucet, about half of the extra carbon dioxide we've put into the atmosphere over the past couple hundred years is still there. It could take several millennia for all of it to be removed. The same goes for any CO_2 we emit this year, next year, and on into the future.

18

Extra CO_2 Going into the Sea Is Making the Ocean More Acidic.

Another name for soda is "carbonated drink." That's because the same carbon dioxide that traps heat in the atmosphere is what soda makers pump into their drinks to make them bubbly. It's not just soda: chances are there was also some CO_2 in the last glass of tap water you drank, though not enough to make it fizz. While most of us don't think about it (maybe because we can't see it happen), gases such as CO_2 can dissolve in liquid on their own, just as sugar or salt can.

This happens all the time in the natural world as gases from the atmosphere dissolve into the oceans. The more CO_2 we add to the atmosphere, the more the oceans absorb. In fact, almost a third of all the CO_2 created by fossil fuels over the past couple centuries has been absorbed by ocean water.

Once it's in the sea, some of this dissolved CO_2 reacts with water to form carbonic acid, which makes the seawater slightly more acidic. The burning of fossil fuels has made the world's oceans about 30 percent more acidic than they were before the Industrial Revolution. Though the oceans are still on the alkaline end of the pH scale, they're moving in the direction of more acidity.

That's bad news for some ocean life, because shells and coral reefs are made almost entirely of calcium carbonate. It's harder to form shells when seawater is more acidic, and it's easier for existing shells to dissolve. There isn't enough acid in the oceans now to destroy shells overnight (and despite a common urban legend, possibly invented by dentists, the acid in Coca-Cola can't dissolve a tooth overnight, either). But there's enough to make it hard for organisms to grow shells. And since tiny shelled organisms are near the bottom of the oceanic food web, acidification could have a ripple effect on the many other sea dwellers that depend on shelled creatures.

Scientists have made progress in understanding exactly how ocean acidification will affect individual species of marine life, and layers of ancient sediments hold some clues about the impact of acidification in the past. Those sediments have been laid down over the ages at the bottom of the ocean, as dirt and dust and the bodies of dead sea creatures form new layers that cover the older ones, year after year and century after century. When scientists dig down to layers that formed during an especially warm period in Earth's history, about fifty-five million years ago, the sediments are unusually dark. They're missing the white remains of shells made from calcium carbonates. The reason, scientists believe, is that ocean acidification during that time caused a massive extinction of bottom-dwelling life, including corals and mollusks. The best evidence to date suggests that today's acidification is happening at an even faster rate than in that earlier time.

19

Cutting Down Forests Means More CO_2 Stays in the Atmosphere.

The burning of fossil fuels such as coal, oil, and natural gas is the main factor behind human-caused climate change, but about 20 percent of the problem comes from an entirely different source: deforestation, or removal of forests. Every year, nearly 80,000 square miles (200,000 square kilometers) of forest are cut down, mostly in the tropics. That adds up to an area about the size of Kansas.

Most plants take in carbon dioxide and use the carbon to form their tissues. The carbon is stored there for the life of the plant. A blade of grass or a giant redwood tree is made largely of carbon, pulled out of the atmosphere during the plant's lifetime. When the plant dies, that carbon is released slowly back into the atmosphere as the plant material rots—unless it's burned, which releases the carbon right away. (For leaves that grow in the spring and fall off in the autumn, of course, the cycle happens over and over.)

When trees in forests are cut down for logging, or fuel, or to clear land for farming, the trees die prematurely, and they die all at once. This has two effects. First, carbon that would normally have remained in "storage" can now be released into the atmosphere. In the case of land clearing for agriculture, the trees are often considered useless waste, and the easiest way to get rid of them is to burn them on the spot. So that extra CO_2 gets pumped into the atmosphere almost immediately. Second, the carbon that would have been absorbed by the trees is now left in the atmosphere.

If all the trees were replaced with new seedlings, those would begin taking carbon out of the atmosphere again. But when trees are cut down for agriculture, they aren't replaced, and the same is often true when it comes to logging. And even if they were replaced, it would take decades—maybe even hundreds of years—for a stand of small, young trees to take in the amount of carbon already released from burning the old, mature stand.

20

Stop All Greenhouse Emissions and the Temperature Will Keep Going Up.

Since 1900, the concentration of CO_2 in the atmosphere has risen from about 300 parts per million to about 390, and simultaneously Earth's average temperature has risen by 1.4°F. Because CO_2 is a heat-trapping gas, climate scientists have little doubt that the two are directly related.

They also know that the relationship between CO_2 and Earth's average temperature isn't a simple one. For one thing, different parts of the climate system (ice sheets, clouds, vegetation, and more) react to warming in ways that can increase the effect of CO_2 alone, or hold it back, in complex ways.

But there's another factor as well: the warming of the planet is lagging behind the growth in greenhouse gases. The reason is that about 70 percent of Earth's surface is covered with oceans and it takes a lot more energy to heat up these huge bodies of water than it does to heat up the atmosphere and the land.

Eventually, the oceans will warm to the point where they aren't absorbing so much of the heat trapped on Earth by greenhouse gases. From that point onward, the heat will go into warming the rest of the planet. When scientists do the calculations, they find that the temperature rise we see today, as CO_2 keeps increasing, is only about half of what it would eventually be if we were to keep atmospheric CO_2 at the current level. Put another way, if humans cut back on emissions just enough to keep CO_2 levels right where they are today, the temperature would eventually settle not at the current 1.4°F or so higher than 1900 but at about 2°F higher. The 90-parts-per-million increase in CO_2 has already committed us to about twice the temperature rise we've seen so far.

That's if we cut back right away. In fact we're not even starting to cut back. We may not even be close to starting. Emissions are continuing to grow. So not only is the temperature going to rise at least a degree higher than it is today; it's going to keep going past that. If we were to stabilize CO_2 concentrations at, say, 450 parts per million by 2050 (which would be a very ambitious goal), the temperature would keep rising for decades afterward. The same goes for 550 parts per million or any other number.

21

Want an Exact Number for How Warm It Will Get? Sorry, Scientists Don't Have One.

It's pretty easy to figure out how carbon dioxide will raise temperatures—in a laboratory experiment, that is. Double the amount of carbon dioxide from where it was in the eighteenth century (about 280 molecules of CO_2 for every million molecules of air, or 280 ppm), and if that's all that changed, you'd raise the temperature by about 1°C, or nearly 2°F.

In the real world, however, things aren't so simple, because when the temperature goes up, all sorts of natural systems react in ways that can drive the temperature up even further (climate scientists call this a positive feedback) or cool things off (a negative feedback).

One example of a positive feedback involves water vapor, which we looked at in chapter 9. When the global temperature goes up, more water evaporates from the oceans. Water vapor is a greenhouse gas, so if CO_2 makes the temperature go up, the extra water vapor that gets added to the atmosphere traps even more heat, raising the temperature even higher.

Another positive feedback, discussed in chapter 4, comes from melting ice, both on the sea (in the Arctic Ocean, for example) and on land (glaciers and ice caps). Ice is white, so it reflects sunlight back into space. When the ice melts, the darker land or ocean underneath becomes exposed. Less sunlight bounces out into space, and more remains to warm the planet.

Still another positive feedback is the melting of permafrost, or permanently frozen ground in Alaska, northern Canada, Siberia, and other frigid places. When rising temperatures thaw the ground, plant matter that's been frozen for centuries starts to rot, releasing additional CO_2 to warm the planet. If the permafrost was originally marshland, it can also release methane, which is an even more potent greenhouse gas.

Methane is also trapped at the bottom of the sea in icy formations known as methane hydrates. If the ocean temperature rises, the methane can escape into the air. A warmer ocean, meanwhile, isn't able to absorb as much CO_2 from the air as a colder ocean can, so however much CO_2 enters the atmosphere from whatever source, more of it will stay there.

All of these are positive feedbacks, but there can be negative feedbacks too. If climate change brings more drought to already-dry areas, winds sweeping through those areas could blow extra dust into the atmosphere. Depending on the size and composition of the dust particles, they could cut down on incoming sunlight, thus creating a cooling effect. The same sort of negative feedback can be caused by aerosols. These are small particles or droplets that are emitted into the air when certain fuels, such as coal or diesel gasoline, are burned.

In some cases, however, it's not immediately obvious whether a feedback will be positive or negative. That's the case with clouds, the biggest source of uncertainty in projecting future temperature. If warming puts more water vapor into the atmosphere, some of it might form extra clouds. Clouds can hold back warming by reflecting sunlight, much as snow or ice does.

On the other hand, they can speed up warming by trapping heat. It all depends on what kinds of clouds they are, how high they are, whether they're over land or water, and all sorts of other things scientists have a hard time capturing in computer models. Most climate models suggest that on balance, clouds will end up adding to the warming, but there isn't enough information yet to say with high confidence what the overall effect of clouds will be. The few studies done so far in the real world suggest that clouds have added some extra warming, but not a lot.

That's not the last of the possible feedbacks, but it's enough to make it clear how complicated the climate really is and how hard it is to come up with precise projections for the future. The balance of evidence right now suggests that a doubling of CO_2 would lead to a temperature rise of between 2.7°F and 9°F, with a most likely value around 5.5°F.

22

Melting Ice Makes the Oceans Rise—but It's Not the Only Factor.

The way most people understand the story about global warming and sea level goes something like this: Earth's atmosphere and oceans are getting warmer, causing ice caps and mountain glaciers to melt. The meltwater then flows into the oceans, causing the water level to rise.

This is true as far as it goes, but there's a lot more to the story. For one thing, water from land-based ice accounts for only part of the rise in sea level over the past century. Most of the rest is due to the fact that water, like most liquids, expands when it's heated. The effect, called thermal expansion, is so small that we don't usually notice it during the course of our everyday lives, except in special circumstances—for example, the liquid that rises in a thermometer when the temperature goes up.

It also turns out that warm water expands faster

than cold water, which means that rising temperatures will cause the oceans to expand faster and faster.

But disappearing ice is becoming an important factor in sea-level rise as well. Sea ice—the kind that covers the Arctic Ocean every winter and that polar bears roam around on—isn't the problem. That ice is already floating, so it doesn't raise sea level when it melts.

Ice that sits on land, however, especially in the massive ice sheets that cover Antarctica and Greenland, does raise sea level when it melts and flows into the ocean—or even when it falls into the ocean without melting. That second process, which is where icebergs come from, has always gone on. Ice sheets are always in motion, spreading outward toward the ocean under their own weight like a thick round of pizza dough under a rolling pin, feeding glaciers that inch toward the sea. But as temperatures have risen, the glaciers have started moving faster.

For a while, scientists thought this was due to melting water pouring down through cracks in the ice to lubricate the area where a glacier's underside meets the bedrock, but that's now looking less convincing. What does seem to be happening is that warming seawater melts tongues of ice where the glacier reaches into the sea, letting the tongues float free from where they had been grinding against ocean floor. When this happens, it's like easing up on the brakes of a car. The glacier moves faster, dumps more ice into the ocean, and raises sea level.

The vast ice sheets that cover Greenland and Antarctica contain a huge amount of frozen water—more than 70 percent of all the freshwater on the planet. As Earth warms, some of that ice is going to melt (although some of that melting may be offset by increased snowfall), and some of it is going to plop into the sea without melting. This process has already begun, mostly in Greenland, where temperatures aren't as cold as in Antarctica to start with.

If all of that ice were to go into the ocean, sea level would rise about 200 feet—20 of it from the icecap covering Greenland, 20 or so from the West Antarctic Ice Sheet, and the other 160 from the much larger and thicker East Antarctic Ice Sheet. Scientists know this partly by measuring how much ice there is. They can also estimate potential sea-level rise by studying times in the distant past when there was pretty much no ice on Earth (it happened during the Cretaceous period, for example, about a hundred million years ago). At those times, sea level was indeed hundreds of feet higher than it is now.

But the only way for all of that ice to melt would be for the temperature to rise quite a lot and then stay high for many thousands of years (how many years depends on how high it goes). Just imagine putting an ice cube into an oven, waiting a few seconds, and pulling it out. In that short a time, even at a very high temperature, the ice cube won't melt much.

Global warming isn't going to raise the temperature as hot as an oven, of course, and the ice sheets are miles wide at their thickest. Even at the high end of where temperatures might go by the end of the century, most of that ice will still be there. Of course, the temperature might continue to climb after the year

2100, which would speed things up. But even then, and even if the temperature stays high, it would take hundreds of years for all the ice to melt.

Nevertheless, scientists believe that if Greenland's temperatures rise more than about 4.5°F and stay there, the melting will be unlikely to stop.

23

Nobody Ever Said Global Warming Means Every Year Will Be Hotter than the Last.

Charts that project how Earth's temperature is expected to rise in the next hundred years sometimes show a smooth line slanting steadily upward. But that smoothness is a simplification. The real world doesn't work quite like that.

You can think of it like a batting average. Baseball fans might be disappointed if their favorite hitter has a bad at bat, but most wouldn't conclude from this that he's getting worse at baseball. There are too many factors that can influence a batter's performance during a single inning—pitching, fatigue, injury, distraction, and of course pure chance. Even a slugger's batting during an entire game doesn't say much about overall ability. What's really important is how well he hits over the course of a string of games or an entire season.

Climate scientists think about climate in the same way. A lot of factors influence Earth's climate, including natural cycles like the El Niño/La Niña currents that warm and cool the Pacific Ocean and the Northern Atlantic Oscillation. These can cause the average global temperature to rise one year and fall the next, but—like a batter having an unusually good or unusually bad inning—these short-term changes don't say much about overall trends.

Global climate over the past decade is a good example. From one year to the next, the changes seem random: average temperature went down from 2000 to 2001, then up for two years, then down in 2004, up in 2005, down in 2006, 2007, and 2008, and up in 2009 and again in 2010. If you made a graph, these temperatures would trace out a wobbly line. But if you zoom out a little bit, you'll see that the wobbly line slants slightly upward over the decade. If you zoom out further, and compare the first decade of the twenty-first century with the 1990s and 1980s, the slant is even more obvious. Each decade is warmer than the one before it.

Even when you look at the graph this way, there are whole decades when temperature doesn't go up. From about 1940 to about 1970, temperatures actually went down a bit. (Scientists think this might have been caused by air pollution that blocked some sunlight. Antipollution laws in the 1970s cleaned up the air and stopped the cooling.) But over the entire twentieth century, the line was still generally upward.

You'll see the same kinds of bumps, including decades with little or no warming, in computer models that project climate into the future. One especially warm or cool year doesn't mean that climate change is getting worse or better; it's always the longer-term trends that tell the real truth.

24

Nobody Ever Said the Whole World Will Warm Up at the Same Rate.

The central fact of global warming is that human-generated greenhouse gases trap energy that would otherwise escape into space, making the planet's average temperature go up. The climate system is always in motion, though, with winds and ocean currents moving heat from one place to another. All of these are likely to change as the planet warms. Beyond that, ice cover and vegetation and clouds and dust in the atmosphere all change as temperatures go up, and these changes lead to faster heating in some places and slower heating in others and even cooling in some places.

For these reasons (and more), scientists don't expect climate change to look the same from one part of the world to the next. Some places are likely to get a lot more snow during the winter, unless it eventually gets too warm for there to be any snow. Other places will get less snow in the winter but more rain in the summer. Some places will get warmer faster; others will get warmer slower. And in any location, there may be more warming in some seasons than in others and more warming at night than during the day—or vice versa.

And some places might have more cold spells, at least during some times of the year. Northern Europe and Siberia have recently seen unusually cold winters, for example. No one is certain why, but one possibility is that melting polar sea ice warms the Arctic and disrupts weather patterns, forcing cold air southward in winter.

Scientists also expect other shifts in wind patterns and ocean currents, which will make some parts of the world heat up faster than average and some parts heat up more slowly. There will also be changes in how much rain and snow different regions receive. Canada, Russia, and other northern countries are likely to get more precipitation year-round during the next century. Parts of North Africa are likely to be drier in both summer and winter. The southwestern United States could transition to permanent drought as a result of drier winters; Australia will get drier summers.

Even if you leave out changes in winds, ocean currents, and precipitation, different parts of the globe respond differently to rising temperatures. Warming at the poles, for example, makes ice begin to melt. As discussed previously in chapters 4 and 21, when the bright white ice melts, less sunlight gets reflected back into space. That means more energy is absorbed by the land and ocean, making temperatures rise even more. If aerosols—soot and other particles—fall on polar ice, the ice gets darker, making it melt faster. Farther from the poles, those same aerosols floating in the air reflect sunlight back into space and have a cooling effect. Those factors contribute to something called Arctic amplification, a trend that scientists have been seeing over the past fifty years. It simply means that the northern polar region is warming faster than the rest of the planet. They expect to see more of it as human-caused warming continues.

25

The Poles Are Warming Faster than Other Places. That's Just What Climate Scientists Predicted.

Lots of factors affect climate in various ways. Depending on which ones are involved—volcanic eruptions, changes in the sun's brightness, shifts in ocean currents, and more—scientists expect to see different patterns in how fast and in what places the atmosphere warms (or cools). Over the past century, and especially the last fifty years, scientists are confident they've detected telltale "fingerprints" on Earth's climate system that point to greenhouse gases (especially CO_2) and aerosols emitted by humans as a factor in climate change. It's very hard to explain the patterns they see by natural causes alone.

One example of a climate fingerprint involves the way that different places on Earth are warming. If, say, shifting ocean currents were solely responsible for the recent warming, the land should have heated up most near the places where warm currents were moving.

Instead, land surfaces are warming everywhere. In fact, they're warming faster than the oceans. So shifting ocean currents aren't the culprit, though they may be a contributing factor.

Scientists have also observed that Earth is warming faster near the poles than it is at the equator. Temperature tends to vary more near the poles naturally, but the current warming would be highly unusual if it were simply a natural variation. If the polar warming were being caused by a brightening sun, for example, things would be reversed: the equator would warm faster than the poles because it receives more direct sunlight.

Another pattern that supports the greenhouse-gas argument is how the atmosphere has warmed over the past fifty years. Over that time, the planet's surface and lower atmosphere have warmed significantly, but the upper atmosphere has actually cooled. That pattern is exactly what scientists expect to see from greenhouse gases because those gases stay in the lower atmosphere and prevent heat from escaping, causing warming below and cooling above. (In this case, greenhouse gases aren't the whole story either: another factor is a drop in ozone in the upper atmosphere.)

This pattern doesn't fit with other causes of warming. When volcanoes erupt, for example, they inject tons of heat-absorbing material into the upper atmosphere, which causes the air high above Earth's surface to heat up while temperatures near the planet's surface drop. (This was observed after Mount Pinatubo in 1992–1993.) And if the sun were getting brighter, it would heat the whole atmosphere evenly, from top to bottom.

Scientists do see some fingerprints from natural forces like volcanic eruptions and shifting ocean currents in the recent climate record. But for at least the last fifty years, man-made causes (primarily greenhouse gases from the burning of fossil fuels) have left a bigger fingerprint than any of those natural forces.

II

WHAT'S ACTUALLY HAPPENING

26

The Atmosphere Now Holds a Record Amount of CO_2—Unless You Go Back Half a Million Years or More.

It doesn't snow much in Antarctica, which is the driest continent on Earth, but since it's also the coldest continent, almost all of the snow that does fall doesn't melt, even in the Antarctic summer. This year's snowfall will cover last year's, next year's will cover this year's, and it goes on and on. The layer of snow that fell a thousand years ago is still there, if you dig down, and the layer that fell a hundred thousand years ago is there too, down deeper. The two-mile-thick slab of ice that covers Antarctica is made up of all of these layers, which are gradually squeezed from fluffy snow into solid ice over the centuries.

The snow is fluffy to begin with because it contains lots of air, and as the flakes are squeezed into ice, the air is preserved in the form of tiny bubbles. Those bubbles are like time capsules holding samples of the atmosphere as it was when the snow fell. The thousand-year-old layer has bubbles with thousand-year-old air. The ten-thousand-year-old layer has ten-thousand-year-old air. Scientists can distinguish one layer from another and date them quite accurately. Looking at the air bubbles tells us how much CO_2 there was in the atmosphere at any given time going back many hundreds of thousands of years.

When they look at the air trapped inside ancient ice, it turns out that over the past 650,000 years or so, CO_2 hasn't risen higher than 300 or so parts per million (meaning that for every million air molecules you count, 300 of them will be carbon dioxide). That includes not only the ice ages but also the warmer periods in between.

When sampling today's atmosphere by contrast (that is, in January 2011; the number will keep rising until and unless CO_2 emissions are reduced), they see CO_2 concentrations of 390 or so parts per million. So there's more heat-trapping CO_2 in the air now than there has been for at least 650,000 years.

27

Sea Level Is Eight Inches Higher than It Was in 1900.

Scientists have been keeping track of sea level since the nineteenth century. At first, they used tide gauges—devices with a float that moves up and down as the water level changes. No single measurement with a tide gauge counts for much if you're trying to measure sea level, since tides and even individual waves make it rise and fall for reasons that have nothing to do with sea-level rise. But if you average the measurements over time, you can get a pretty good idea of where sea level actually is.

More recently, scientists have used satellites that measure their own height above the sea by bouncing radar or laser beams off the sea surface and seeing how long it takes the beams to return. Again, due to waves, no single measurement means much, but when you average thousands of measurements, the result is even more accurate than what you get from tide gauges.

Overall, these measurements show that sea level as

of 2009 was about eight inches higher than it was in 1880. There's also some evidence that the rise has been faster lately: the average rate of sea-level rise from 1961 to 2009 was about 0.07 inches per year. But if you look just at the years 1993–2009, the rate was nearly double that, or about 0.12 inches per year. The increase could be due to a stronger influence of global warming, since warming melts ice and makes seawater expand.

It could also come from natural forces, but those don't explain the longer-term increase since 1900.

Finally, it's important to note that this is an average rate of sea-level rise for the entire world. Locally, the rate can be higher or lower, since the land itself can be slowly rising or falling due to geologic forces. Changes in ocean currents can also make sea level locally higher or lower than the global average.

28

Earth's Temperature Is About 1.4°F Higher than It Was in 1900.

This seems like a small number. It's so small, in fact, that it might not sound especially alarming. After all, we experience much bigger changes in temperature every day. If the room where you are sitting heated up by 1.4°F in the next five minutes, you wouldn't even notice. In most places on Earth, the temperature changes by more than 1.4°F over the course of a couple hours and by at least ten times that amount from day to night.

But it doesn't make much sense to compare our daily experience of temperature change with the global change of 1.4°F. First of all, daily ups and downs tend to even out in the long term. They don't say anything about whether the planet as a whole is changing.

Beyond that, your daily experience is local, whereas Earth's average temperature is global. Daily changes, and even heat and cold waves, come from warm or cold air and water moving from one part of Earth to another. That happens all the time, as winds change and ocean currents shift. But the 1.4°F global increase since 1900 means that the planet as a whole has warmed, which means that it has been absorbing more heat than it's getting rid of. Earth has an energy imbalance.

Although greenhouse gases have been steadily increasing, temperatures have not increased in the same manner. In fact, most of the warming happened after 1979. Toward the beginning of the century there was a period of slow warming. Then, in the middle of the century, global temperatures were relatively stable. They even declined a little. The ups and downs over the first half of the century show that natural forces like volcanic activity and changes in the sun can have a significant effect on global temperatures—more, during that period, than greenhouse gases from human activities. If greenhouse gases were the only driver of temperature in the first part of the century, temperatures would have risen steadily.

By the late 1970s, the effects of greenhouse gases began to overshadow natural forces. Despite natural cooling effects, like the aerosols emitted in the eruption of Mount Pinatubo and other volcanoes, the global thermometer has risen significantly over the past several decades. The 1980s were warmer than any previous decade in the twentieth century. The 1990s were even warmer, and the first decade of the twenty-first century was warmer still. On average, the planet's temperature rose by about 0.8°—over half of the twentieth century's overall warming—between 1979 and 2005.

29

The Continental United States Had Twice as Many Record-High Temperatures During the First Decade of the Twenty-First Century as Record Lows.

Just because a particular day is the warmest on record doesn't mean the planet is warming. Just because a particular day is the coldest on record doesn't mean it's cooling. That's because the temperature for a single day is governed by all sorts of factors, including atmospheric pressure, cloud cover, and the direction of high-altitude winds that drive weather systems.

It's only over decades that true patterns emerge to show that Earth might be warming or cooling. One way to see those patterns is to look at average temperatures around the world, month after month and year after year.

But you can also see the patterns in record highs and lows—as long as you look at enough of them at once. If Earth is warming, you'll still see record lows for particular dates (again, because a single day's tem-

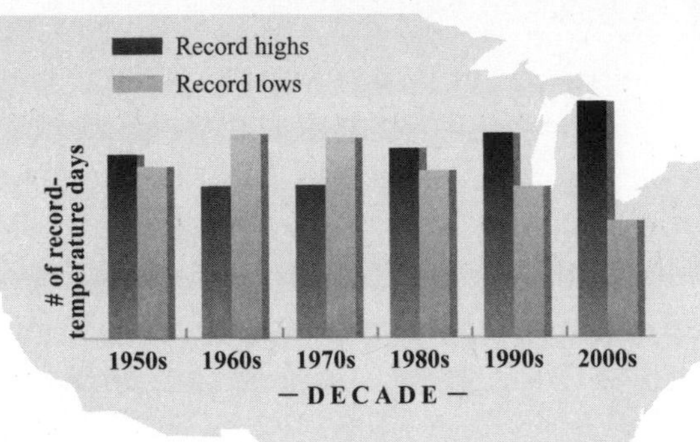

perature has so many influences), but you should see more record highs.

That's just what scientists see in the continental United States. In the 1960s, there were more record lows than record highs. It was the same in the 1970s. But in the 1980s, there were about 14 percent more record highs than lows. In the 1990s, there were 36 percent more record highs. And in the first decade of the twenty-first century, there were more than twice as many record highs. By 2100, according to projections by computer models, record highs could conceivably be fifty times as common in the United States as record lows, if greenhouse-gas emissions continue to grow at the current rate.

30

Glaciers and Ice Caps Have Been Shrinking Since About 1850.

Over the last century and a half, people have been keeping pretty good track of the changing shape and size of glaciers in the European Alps. Together, those records add up to a decisive trend: the glaciers are shrinking. The Alps have lost about half their ice since 1850. Part of that loss has been linked to natural variations in climate, like shifting ocean currents that warm and cool the region over decades.

But natural variations don't explain all of the melting in the Alps or in other parts of the world. Mountain glaciers all over the planet are shrinking faster than they were a century ago. There was a slow decline in glaciers between 1850 and 1950, followed by a short growth spurt. But since 1970, the world's glaciers have been shrinking at an accelerating rate.

Of course, this is an overall trend. It doesn't mean that every glacier on every mountain has been losing ice faster with each passing year, or even that every glacier has been losing ice. Glaciers grow or shrink depending on how much snow they add and how much ice they lose over a given amount of time. Warmer air can hold more water, which can lead to more snow in winter, so glaciers can actually grow as temperatures rise. In Norway in the 1990s, for example, air temperatures were warming, but there was so much snowfall that the glaciers got bigger.

On the whole, the melting side of the mass-balance equation has been winning out. Over the last couple of decades the most melting has occurred in Alaska, the northwestern United States, Canada, and Patagonia.

Scientists expect the melting trend to continue through the twenty-first century. Smaller glaciers are particularly susceptible to increasing temperatures (the same way that small ice cubes melt faster than big ones). If the melt continues to accelerate, many small glaciers will be virtually wiped out by 2100.

31

Greenland Is Losing Ice Faster All the Time.

Greenland isn't quite as huge as it might seem, since many world maps distort the size of landmasses in the Arctic and Antarctic. Still, it's the largest island in the world, more than three times the size of Texas and about two-thirds as big as India. More than 80 percent of that vast area is buried under a massive ice sheet, which is about two miles thick at its deepest part.

Greenland's ice is constantly renewed by fresh snow every year. It's also constantly shrinking as glaciers flow slowly from the center out toward the sea. The ends of many glaciers eventually melt or break off as icebergs. (The iceberg that sank the *Titanic* in 1912 probably came from Greenland.) Greenland also loses ice through melting.

If the snow falls faster than ice can melt and icebergs can break off, Greenland's ice sheet gets bigger. If the combination of breakage and melting is faster than snow can build up, then the ice sheet gets smaller. Right now, the ice sheet is clearly getting smaller. In fact, one recent study puts the shrinkage at about 105 billion tons of ice per year.

This is a sharp acceleration from what was happening in the past: as recently as the 1990s, Greenland was losing only about 7 billion tons per year. Now those hundreds of billions of tons of ice are all going into the ocean, either as water or as icebergs, and contributing substantially to sea-level rise. If all of Greenland's ice melted or fell into the sea, it would raise sea level by over twenty feet, but that's not likely to happen for many hundreds of years at least.

There's no mystery about why the ice is melting. Since the early 1990s, air temperatures in Greenland have risen about 4°F on average. In summer, that's enough to melt ice from the top of the ice sheet. At

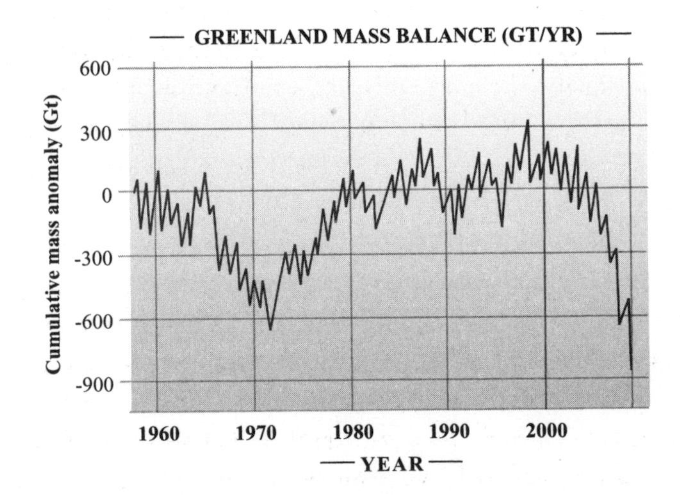

the same time, warmer ocean temperatures are melting some of the ice down where the glaciers meet the sea. With that ice gone, the rest of the glacier has less to push against, so that helps speed up the glacier's motion as well.

One way scientists measure ice loss is with satellites like GRACE (short for Gravity Recovery and Climate Experiment). It's really a pair of satellites that measure how gravity varies across the surface of Earth. Mountain ranges, for example, have lots of extra rock, so there's extra gravity. Deep valleys, on the other hand, are mostly empty space, so there's less gravity than average. Greenland is more or less a giant plateau of ice, so it has extra gravity, but as the ice shrinks, GRACE detects a decrease in gravity, which lets it measure the total loss of ice.

Another satellite, called ICESat, measures the growth or shrinkage of the ice sheet by sending a laser beam down to bounce off the surface. If the beam takes longer to make the round-trip from satellite to ice and then back to the satellite this year than it did last year, that means the surface is farther away; the ice has gotten thinner. Right now, it's getting thinner at the edges of Greenland as glaciers are flowing faster and stretching thinner to reach the sea.

One thing complicates all of these measurements, however. At the end of the most recent Ice Age, the melting of ice sheets in North America and Europe lifted a huge weight off those continents. Imagine Earth as a giant balloon, and the ice sheets were like hands pressing down on part of that balloon. When the hands lift off, the balloon wants to bounce back.

Similarly, with the ice sheets gone, Earth wants to bounce back to its original shape. This takes time, however. Even though the ice sheets disappeared ten thousand years ago, the land is still inching its way upward in relief. Places where there were no ice sheets—the bottom of the sea, for example—are sinking, to compensate.

According to one study, Greenland itself is sinking along with the seafloor that surrounds it. If other studies continue to find the same thing, it means that some of Greenland's shrinkage (though not all) comes from the underlying land, not melting ice. The figure cited above of 105 billion tons of ice lost per year is lower than many previous estimates because it takes into account the sinking of the Greenland landmass.

32

Polar Bears Will Suffer as Sea Ice Continues to Melt.

For many people, polar bears have become the most familiar symbol of global warming. In 2006, to name just one example, *Time* magazine featured a polar bear on its cover, along with the words "Be Worried. Be Very Worried." In 2008, the bears became the first animal to be listed as threatened under the Endangered Species Act because of climate change. "Threatened" means a species is at risk of becoming endangered, which in turn would mean that it is on the brink of extinction.

Normally, the U.S. Department of the Interior makes these decisions based on clear evidence that populations of the species are shrinking in response to human activities such as habitat destruction, poaching, and pollution.

But with polar bears, it's a slightly different story. The decision to list polar bears was based more on expectations about the future than on hard data about declining populations today. Biologists know that polar bears depend on sea ice for hunting, for fishing, and sometimes for building dens. Scientists who study sea ice know that since the late 1970s, summer sea ice has been declining in the Arctic Ocean. If people continue to emit greenhouse gases at the current rate, climate models predict that the Arctic could be virtually ice-free in late summer, maybe as early as 2050. Biologists estimate that those conditions will result in the extinction of two-thirds of the world's polar bears. Late in the first decade of the twenty-first century, ice was melting faster than models projected. If that trend continues, the ice-free date could come even sooner.

Although the bears' habitat is shrinking, it's not clear whether polar bear numbers have started to decline in response. Experts estimate that there are somewhere between twenty thousand and twenty-five thousand of the huge white predators alive today, spread across the world's northernmost latitudes. But because the bears' range includes some of the most remote, inaccessible territory on the planet, nobody really knows if that number is higher or lower than it was, say, fifty years ago.

What researchers do know is that there are nineteen distinct populations of polar bears. Eight of those are currently in decline—partly due to loss of sea ice, but also due to hunting and environmental pollutants. In Canada's western Hudson Bay, there's been a 22 percent decline in the local polar bear population since the 1980s. But the population in south Hudson Bay, as well as two other populations, has been stable. One population is increasing. As for the other seven populations, there isn't enough information to say anything for sure.

Because we don't actually know that polar bears are on the decline worldwide, the "threatened" listing has been controversial. Environmentalists often use polar bears as a mascot for climate-change activism. Summer melting has forced some polar bears to swim very long distances without finding ice floes to rest on, and some have drowned. Some of these drownings have been widely reported in the media. But critics argue that it's just an attempt to play on people's emotions, without hard numbers to back up the implication that drownings are happening more often than they used to.

It's clear, however, that as sea ice continues to shrink, loss of habitat will make it harder for polar bears to survive. When they're forced to spend more time on land, the bears are cut off from their energy-rich food source: ocean-dwelling seals. Though they are capable of expanding their diet to all kinds of foods if they have to—eggs, mosses, trash—those alternatives don't offer anywhere near the amount of nourishment that seals do. In the western Hudson Bay, female polar bears are already slimming down due to calorie cuts resulting from the loss of their hunting grounds. Reproduction rates are falling as a result. Perhaps just as important, as polar bears spend more time on land, they will run into people more often. Although these encounters can be detrimental to people, it is the polar bears that will lose in the end.

Polar bears are very unlikely to be winners in a warming climate.

33

The Growing Season in the Continental United States Is Two Weeks Longer than It Was in 1900.

The official first day of each season is determined by the position of Earth in its orbit around the sun. Since that doesn't change much from one year to the next, the official dates don't change by more than a day or two either. The growing season is different: it starts whenever conditions for plant growth—including temperature and sunlight—are right. Since freezing temperatures can damage or destroy crops, the growing season is often defined as the stretch of time between spring and fall when temperatures stay above 32°F around the clock.

Over many temperate regions of the world, studies have shown that this continuous stretch of frost-free time has been getting longer. During the twentieth century, the growing season in the lower forty-eight United States lengthened by an average of two weeks. This trend toward longer growing seasons within the

United States is happening just about everywhere, but the change has been much greater in the West than in the East. In fact, the western United States has gained an average of nineteen days, while the growing season in the East lengthened by just three days.

On a smaller scale, there are some more extreme examples: in the Sonoma/Napa region of California, for example, the growing season has lengthened by a full 66 days, from 254 to 320, since 1950—much to the delight of winemakers in the region. (Longer periods without frost aren't a good thing in all cases, however; frost can, for example, keep insect pests under control. And many plants—for example, winter wheat—require cool winter temperatures during their life cycles.)

Natural variations in climate account for some of these changes, especially during the first half of the twentieth century. During that time, trends in annual air temperatures match up pretty closely to the length of the growing season: colder decades had shorter growing seasons, and warmer decades had longer ones. More recently, the trend has accelerated, with half of those extra frost-free days added since 1980. During that time the growing season lengthened faster than average annual temperatures increased.

Here's why: though annual average temperatures affect the length of the growing season, other factors matter a lot too, like the difference between daytime and nighttime temperatures. In recent decades, that difference has gotten smaller. This is just what scientists expect to see as greenhouse gases in the atmosphere increase: the heat they trap keeps the surface warm even when the sun isn't shining. If the thermometer stays just a little above freezing at night, it doesn't affect average overall temperatures very much, but it does have an impact on how many frost-free days there are. Studies have clearly shown that these changes in the growing season worldwide are linked to warming from greenhouse gases during the second half of the twentieth century.

The earlier spring and later fall have led to other changes as well. In the West, there's an annual "pulse" of water every spring, as accumulated winter snow begins to melt and fills up streams and rivers. This pulse came seven to ten days earlier, on average, in the second half of the twentieth century. Earlier thaw has also led to changes in the blooming times of flowers, and warmer spring temperatures have changed migration, breeding, and nesting timing of several insects, amphibians, birds, and mammals.

Much of the life on Earth runs according to the seasons. In temperate climates during spring, the ground thaws, seeds sprout, grass grows, birds nest, and babies are born; in autumn, the leaves turn, fur thickens, birds reverse their spring migrations, bats hibernate. People have been keeping track of the timing of some of these life events for a long time. One of the longest records is the spring flowering date for cherry trees in Kyoto, Japan, which goes back twelve hundred years.

Those long-term records reveal that while the same events happen year after year, they don't generally happen on exactly the same date. Over the centuries, for example, the cherry trees in Kyoto have blossomed as early as March and as late as May. Changes in Earth's position around the sun—the reason we have seasons—happen every year like clockwork. But the timing of life events (biologists call this phenology) is based on factors—like climate—that vary across years, decades, and centuries.

Natural variations in climate cause plants and animals to adjust their annual calendars backward and forward, but the warming trend over the twentieth century is pushing them mostly in one direction: life events have been happening earlier in the spring. The twelve-hundred-year record of cherry trees shows no discernible trend until the mid-nineteenth century, when the flowering dates began to advance. In the 1980s and 1990s, average flowering times became earlier than previously recorded.

It isn't just cherry blossoms that are getting an early start. Thousands of plants are budding and fruiting earlier. Frogs and mammals are breeding earlier. Butterflies are making their annual migrations earlier. Birds are building their nests earlier than they were a few decades ago. In 2010, scientists in the United Kingdom studied records of over seven hundred species from 1976 to 2005 and found that 80 percent of the species had moved up their spring calendars. Other major studies have shown the same trends occurring all over the world.

Though the overall pattern is the same—life events occurring earlier each decade—for all kinds of organisms, they aren't all moving at the same pace. Across the Northern Hemisphere, amphibians are breeding an average of over seven and a half days earlier every ten years, while the first flight of butterflies moves up by a little over three and a half days, and herbs, grasses, and shrubs send out their first shoots about one day earlier per decade.

There's also a lot of variation within similar groups of organisms. In the United Kingdom, for example, bird species that migrate only short distances are returning to their summer quarters earlier than birds that travel farther. For at least one bird species—the cuckoo—this difference is causing problems. Cuckoos are long-distance migrants, traveling all the way from tropical Africa to the United Kingdom in the spring to breed. They don't build nests when they arrive, though, as other birds do. Instead, female cuckoos sneak their eggs into strangers' nests and let the unwitting hosts do all of the work raising their chicks. As short-distance migrants arrive earlier, their nest building and breeding begin earlier too. By the time the cuckoo arrives on the scene, looking for nests to sneak into, it's sometimes too late: some of the hosts have already hatched their young.

That's just one of the problems caused by earlier springs. Other mismatches between species are beginning to disrupt the relationships between predators and prey, though scientists don't know what the long-term consequences of those disruptions will be.

34

Ecosystems Around the World Are Already Seeing Big Changes as the Climate Warms.

People often think of ecosystems in terms of the plants and animals that live in them, but a lot of other factors, like geology, altitude, and climate, are crucial to determining what kind of ecosystem develops in a particular place. When one of those nonliving factors changes, ecosystems change too.

Over the twentieth century, global temperatures increased by an average of about 1.4°F, but some places have warmed a lot more than this, and other places have warmed less. These temperature increases have been enough to trigger changes in ecosystems all over the world, especially in places where the warming has been the greatest. In some places, the changes have been subtle, like a slight shift in vegetation that only a careful observer would notice. In other cases, small changes in climate have sparked a chain of larger effects, leading to massive changes.

The biggest climate-caused ecosystem shifts today are happening at the world's most northern latitudes, where the temperature over the last century has been rising about two times faster than the global average. In Alaska, for example, warming has paved the way for a spike in the numbers of spruce bark beetles. Bark beetles have been a pest to Alaskan white spruce for thousands of years, but their numbers were held in check by the cold climate, which forced the insects to hole up in the bark of individual trees for most of the year. As the length of the warm season increased over the 1980s and 1990s, however, bark beetles had more time to fly from one tree to the next, burrow, and lay their eggs between the bark and the wood. The beetles had another thing going for them, too: a multi-year drought had weakened many of the spruce trees, leaving them vulnerable to attack. In the mid-1990s, the bark beetle population exploded, and over the next few years the pests wiped out white spruce forests over an area the size of Connecticut. In the years since, the combined forces of a longer insect-breeding season and forest management practices that left forests overcrowded gave way to similar epidemics farther south. Large swaths of pine and spruce have been destroyed by insects in New Mexico, Colorado, Wyoming, and elsewhere.

In the late 1990s, the effects of the bark beetle epidemic rippled throughout Alaska's white spruce ecosystem and affected virtually every population of living organisms, but not all of the impacts were negative. Fewer spruce trees meant a sunnier understory in the forest, which allowed grasses to move in and take hold. The grasses, in turn, changed the soil temperature, making the environment more friendly for some other types of vegetation. Animals that feed on grasses, including moose, elk, and some birds, also benefited.

But the beetle infestation was bad news for organisms that rely on white spruce for their habitat, like hawks, owls, red squirrels, and voles. Voles—a type of small, round rodent—are an especially vital part of the ecosystem because they help spread mycorrhizal fungi, which attaches to the roots of plants and helps them take in water and nutrients. Voles are also an important food for a number of predators.

Since ecosystem changes always hurt some living creatures and help others, it's hard to say whether a change is good or bad overall. Instead, people who study ecosystems often focus on the impacts on a single species: for instance, us. In the short term, the Alaskan spruce beetle epidemic supplied a lot of people with firewood, but only by destroying tons of otherwise valuable timber and threatening the livelihoods of loggers. And no one knows for sure what the long-term impacts on the forest will be. Ecosystems tend to return to their previous states after disturbances like pest outbreaks, fires, or major storm events, but if the Alaskan spruce ecosystem is disturbed too often or too much, it might shift to a different type of forest, a woodland, or a grassland instead.

In extreme cases, major assaults on ecosystems can lead to a total collapse in which the ecosystem doesn't bounce back to the way it was or transition to a new, healthy state. The result is an area with very little life; in the oceans, biologists refer to these areas as dead zones. One example of ecosystem collapse is the coral reef die-off that happened in the Indian Ocean in the late 1990s. The reefs had been up against multiple pressures for several decades, including reduced populations of fish (a result of overfishing), ocean acidification due to increased CO_2, and rising water temperatures. Then, in 1998, a strong El Niño event moved warmer-than-usual water into the coral reefs, causing massive coral bleaching. As a result, one-quarter of the coral in the Indian Ocean died, taking with it much of the marine life that depended on it for survival. Eight years later, researchers reported that in some places in the Indian Ocean, the number of fish species has been reduced by half. Other coral reef collapses occurred in parts of the Caribbean in 2005 and 2010, also due to extreme heat events.

35

Some Species Can Adapt to Changing Climate a Lot Better than Others.

The fennec fox, which lives in the Sahara desert, is in the same family of mammals as the arctic fox, but it has evolved to cope with a very different sort of environment. Arctic foxes have thick white fur, which helps them stay camouflaged during the snowy winter. During the springtime, arctic foxes shed some of that fur to avoid overheating. What's left changes color from white to brown to maintain the camouflage as the snow disappears.

The fennec fox, by contrast, has light brown fluffy fur, which reflects sunlight during the day but keeps the animal warm when temperatures drop at night. And unlike the arctic fox, the fennec fox has enormous ears that radiate heat and help it stay cool. Each fox is so well adapted to its specific climate, in fact, that it probably wouldn't be able to survive if the conditions suddenly changed.

The same holds true for most organisms in the Arctic and the Sahara—but not necessarily all of them. Some plants and animals are more specialized than others. If the entire ecosystems of the fennec fox and the arctic fox were to switch places, some organisms would undoubtedly die in a matter of hours. Others would languish over time. And a few lucky ones, with the ability to tolerate a wide range in temperatures, might actually survive and reproduce.

That's just a thought experiment, of course; whole ecosystems don't get picked up and set down in a new climate. But when climate change makes the conditions of an ecosystem unfavorable to a particular species, the population begins to shift. If temperature rises, for example, some of the organisms at the southern, warmest edge of the ecosystem may suffer and fail to reproduce, while some organisms at the northern, coolest edge reproduce and migrate a little farther north. Biologists first noticed this effect hundreds of years ago. In Europe, records from the eighteenth and nineteenth centuries show that populations of birds and butterflies shifted north during warmer decades and shifted south during cooler decades.

Since the 1960s, rising temperatures have pushed the home range of hundreds of species in two directions: northward (in the Northern Hemisphere, anyway) to higher latitudes or uphill to higher elevations. In Europe and North America, for example, populations of thirty-nine butterfly species moved up to 125 miles north over a period of twenty-seven years. Cold-loving plants in the Alps moved higher into the mountains. Sea anemones off the California coast moved northward. Lowland birds in Costa Rica extended their range to higher altitudes.

But because each type of plant or animal within a given ecosystem has a different tolerance range, each will respond in its own way to the change. While some species in an ecosystem are on the move, others may not move in the same direction or the same speed, or they may not even move at all. The result over time is that the set of interrelated organisms we think of as a cohesive "ecosystem," with all its complex predator-prey relationships, begins to break apart. Species that have been living together gradually separate and start living with new species.

Fossil evidence from the last million years shows that ecosystems that existed under different climates than we have today had mixes of species that were different from those in modern ecosystems. As temperatures continue to rise and different organisms continue to move to find their preferred habitats, scientists expect some current ecosystems to break apart and new ecosystems to form.

36

The Arctic Has Been Losing Ice Much Faster than the Antarctic. That's Just What Scientists Expected.

Summer ice in the Arctic Ocean has been in steep decline over the past several decades. During that same period, however, the ice that covers the sea around the edges of Antarctica in winter has been increasing slightly. At first, this seems to make no sense. How could global warming be happening at one end of the planet and not the other?

The answer, it turns out, is that sea ice has been increasing even though Antarctica itself, and the Southern Ocean that surrounds it, have been warming. One reason for the extra ice growth is the "ozone hole," caused by chemical pollution and first discovered in the atmosphere above the South Pole in the 1980s. The loss of ozone has changed wind patterns around Antarctica in such a way that existing sea ice has been dispersed, creating large areas of open water where new ice can easily form. The ocean is warmer than it used to be, but it's still plenty cold.

The other part of the story has to do with an increase in rainfall and snowfall over recent decades. This is happening because in a warmer world, more water evaporates from the ocean and eventually falls again. In the Southern Ocean, this forms a layer of freshwater on the surface, and freshwater is easier to freeze than salt water. The freshwater layer keeps the relatively warm ocean water underneath from rising. The result is more ice.

Finally, it's important to note that while the sea ice around the rim of the Antarctic continent has been growing, the massive ice sheet that sits on the continent itself has not. At least not since 2002, when a satellite first started tracking changes in the ice cap. In fact, the world's most massive ice sheet has been shrinking, and it seems to be shrinking faster all the time. Between 2002 and 2006 the ice sheet lost an average of 104 billion tons of mass per year. But between 2006 and 2009 that rate more than doubled, to 246 billion tons per year.

So while the growth of Antarctic sea ice shows how complex the science of climate change can be, it doesn't change the fact that there's still a lot of ice melting.

37

Arctic Sea Ice Has Been on a Mostly Downward Spiral for the Past Thirty Years.

Every summer above the Arctic Circle, the sun rises relatively high in the sky and stays up for as much as twenty-four hours every day. Ice on the surface of the Arctic Ocean melts back, exposing open water. But the Arctic is still cold enough, even in summer, and the ice is thick enough that it has never melted away completely, unless you go back many hundreds of thousands or even millions of years.

As global temperatures rise, however, scientists expect to see more open water each summer, on average. It has to be "on average" because wind patterns, weather systems, and water currents can have a bigger effect than global temperature in any given year. (Besides, even though global temperatures are rising steadily, they don't change by the same amount each year.) That could mean an unusually large or unusually small amount of melting in one summer.

Over time, though, the melting should increase. That's not just because of rising temperatures. As we've seen, bright white ice reflects a lot of sunlight back into space, so this "lost" energy can't heat the surface. But when the ice melts, it exposes water, which is much darker and doesn't reflect nearly as much light (it's almost black). More heat stays on the surface, so air temperature goes up and causes even more melting. This feedback effect—more melting means less reflection means more heat means more melting means less reflection and so on—is called Arctic amplification, and it makes the Arctic warm up faster than the global average.

Sure enough, Arctic sea ice has melted more each summer (on average!) than the summer before since the late 1970s; and in fact, the downward trend in late-summer ice is steeper than nearly all climate models predicted. It's possible that the trend started even earlier than that, but until about thirty years ago scientists didn't have satellites orbiting overhead to survey the entire Arctic. They mostly had to use ships and planes, which can only make limited observations.

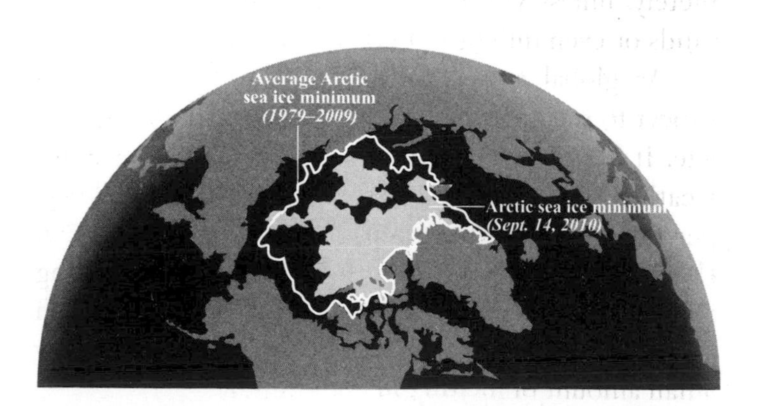

Over the years 1979–2000, the amount of ocean covered by ice in September—the time of year when there's the least ice, thanks to a whole summer of heating—averaged about 7 million square kilometers. In recent years, it's been well under 6 million, and in 2007, the lowest year on record so far, it was about 4.2 million (in 2010, it was just under 4.8 million). In recent years, including 2010, both the Northwest Passage along the northern coast of Canada and the Northeast Passage (or Northern Sea Route) along northern Siberia were open at the same time—something that has rarely happened in the past. Arctic sea ice in the winter of 2010–2011 covered less total area than every previous year on record.

If melting trends continue as they have in the recent past, the Arctic Ocean could be almost completely ice-free, for at least part of the summer, by 2100. But there's one more factor to consider. When open water in the Arctic refreezes over the winter, it turns into a relatively thin layer of what's called, for obvious reasons, first-year ice. The ice that survives summer without melting will often add more ice over the following winter. Multiyear ice (which can add layer after layer over the years) is, naturally, thicker than first-year ice, so come summer, it's harder to melt.

As the Arctic has seen more and more ice-free open water in recent summers, there's more thin, first-year ice and less thick, multiyear ice. From 1980 to 2005, for example, the average thickness of Arctic ice in fall and winter dropped by more than a third, and from 1980 to 2008 winter thickness decreased by nearly half.

This means that when an especially warm summer

comes along, or conditions change in some other way, a bigger expanse of ice is prone to sudden melting. If that happens, Arctic amplification could rapidly create a situation where the Arctic Ocean becomes ice-free much sooner than 2100—some predictions put it at 2050 or even earlier.

But while that's a possible scenario, it also remains a theoretical one. An ice-free Arctic Ocean in summer is likely to come along eventually, but it's hard to say when. Once it does happen, the extra heat absorbed by the ocean, plus the fact that all of the ice formed afterward will be thin, first-year ice, makes it likely that the Arctic Ocean could be more or less permanently ice-free in summer.

Even so, the Arctic Ocean will probably always have ice during the winter months. This is an example of a larger point: low temperatures, cold air, ice, and snow will continue to be present on Earth even in a warmer overall climate—just less prevalent than they are now.

38

Droughts, Torrential Rains, and Other Extreme Weather Are Happening More Often than They Used To.

Based on their understanding of how the climate system works, scientists expect that a warming Earth will see more and more episodes of weather extremes such as droughts, floods, heat waves, and severe storms. The reason for more heat waves is pretty obvious. If you define a heat wave as an extended stretch of days when the temperature is well above what we now think of as normal in a month when temperatures are already high (July and August in the United States, for example), then a generally warmer world will have more stretches like this.

With droughts and floods, the reason may be less obvious. In a warmer world, more water evaporates from the oceans and other bodies of water and also from the soil. Land will tend to be drier, but when the rain or snow does come, there's more water vapor stored up in the atmosphere, so precipitation should become heavier. That in turn could lead to a higher risk of flooding. Atmospheric scientists calculate that the warming over the twentieth century should already have put about 5 percent more water vapor into the atmosphere today, on average, than there was in 1900.

Even so, it's difficult to prove definitively that extreme weather events have increased, unless we're talking about averages over the entire globe, or at least over very large regions. That's because the most extreme events—the worst storms, droughts, and floods—tend to come less often than ordinary events. Climate and weather scientists talk about "hundred-year floods," for example, meaning a flood so big it should happen about once every hundred years (or ten times in a thousand years), on average, in a particular location.

But that's only an average, so if you have two such floods in a hundred-year period, does that mean the risk has doubled, or maybe that you are now "safe" for the next two hundred years? Or is it just a statistical fluke, similar to flipping a coin and getting two heads in a row just by chance? It can be hard to tell. It's even harder when you don't have good measurements over the past century in all parts of the world, and when different parts of the world respond to climate change in their own ways.

Even today, with much better monitoring, it's a challenge for scientists to gather enough information to draw firm conclusions about long-term trends over large regions. It's more difficult still to say anything concrete about the role of climate change in single events. No single event, such as the deadly heat wave that hit Europe in 2003 or the 2010 floods in Pakistan or heat and wildfires in Russia, can be definitely blamed on climate change: we think that climate change has made events like these more likely, but that doesn't mean they couldn't have happened anyway.

For single events like these, the best scientists can do is calculate whether climate change has tipped the odds in favor of such events happening—whether the coin you're flipping has become slightly loaded in favor of tails, for example. A study published in 2005 showed just that kind of loading for the European heat wave of 2003.

As for general trends, here's what the Intergovernmental Panel on Climate Change (IPCC) said about extreme weather: "Globally, the area affected by drought has likely increased since the 1970s. . . . It is likely that heat waves have become more frequent over most land areas. . . . It is likely that the frequency of heavy precipitation events . . . has increased over most areas. It is likely that the incidence of extreme high sea level has increased at a broad range of sites worldwide since 1975."

The word "likely" in the IPCC report means that there's at least a two-thirds probability that something is true—meaning that while it's higher than fifty-fifty, there's still a sizable uncertainty, for the reasons explained above. What scientists can say is that events like this should happen more often in a warmer world. If our planet continues to warm as scientists expect—and as our record of observations gets longer and more comprehensive—scientists will be able to make more definitive statements.

But they're also trying to do better with the information they have today. Using special statistical techniques designed to look at extreme events, scientists have done studies, published after the last major IPCC report on global climate, that look at changes in extreme temperature and precipitation events over large areas of the world. They've detected evidencc of a human role in these changes. However, it's still very hard to say anything definitive about the human fingerprint on specific, local types of extreme weather.

39

Rising Ocean Temperatures Are Causing a Major Die-Off in Corals.

Coral reefs support some of the most diverse and thriving ecosystems in the world's oceans. They're created by colonies of small sea organisms that surround themselves with shells made of calcium carbonate, the same material clam and oyster shells are made from. Over time, if conditions like water temperature and salt content are right, those colonies of coral can grow into reefs.

Each coral has tiny algae living inside of it—up to several million algal cells in an area of coral the size of a fingernail. Usually, this arrangement works out well for both organisms. The corals offer the algae nutrients and a nice place to stay, away from predators, and the algae provide the corals with sugar from photosynthesis.

When the water temperature gets too high, though, the coral reacts by evicting the algae. That's deadly for the algae, but without the sugar it's come to rely on, the coral can eventually die too. The loss of algae turns coral from its usual brown to white, the natural color of the coral's shell. For that reason, the process is known as coral bleaching.

Coral bleaching can and does happen naturally due to a variety of environmental shocks, like sudden changes in temperature, salt concentration, or light. But the overall increase in ocean temperatures due to greenhouse warming is causing it to happen more often. Since the surface of the oceans is huge and heats up slowly, the total increase in water temperature over the past century is just a small fraction of the increase in air temperature. But even that small increase can be enough to start coral bleaching.

Coral reefs can actually survive bleaching, just as they can recover from hurricanes or other destructive events. But as is the case for all living things, their ability to survive and recover depends on how healthy they were to begin with. Today, many of the world's coral reefs are already in trouble due to pollution, overfishing, and physical destruction by tourists and fishermen.

It's a combination of all of those factors, including rising temperature, that has destroyed 30 percent of the world's warm-water reefs since 1980. But as more heat is trapped by greenhouse gases in Earth's atmosphere, and as more CO_2 in the oceans makes seawater more acidic, carbon emissions could play a bigger role in the destruction of corals in the future.

III

WHAT'S LIKELY TO HAPPEN IN THE FUTURE

40

Computer Models Aren't Perfect. This Isn't a Big Surprise.

Computer models simulate things in the real world. Engineers use them to design cars, airplanes, and other things we depend on every day. Climate models use basic scientific equations about energy and matter to simulate the planet's atmosphere, oceans, land ice, sea ice, and vegetation coverage. Climate scientists use the models to help them understand how our climate works, how it behaved in the past, and how humans are affecting it. They also use them to explore what the climate might be like decades from now.

Of course, there's no way to test these predictions directly (unless we wait for decades, by which time we wouldn't need them anymore). But there are still ways to figure out how well the models work, because models aren't just used to look at future changes.

Models can be tested, for example, to see how accurately they simulate today's climate—how closely they match up with the average patterns of temperature or precipitation or air pressure that we observe in the real world.

Scientists also test their models by simulating past climate changes, say, between 1850 and today. When they take into account natural variations in the sun's intensity, aerosols (that is, tiny particles) from volcanic eruptions and industrial pollution, and greenhouse gases emitted by humans, the models reproduce the rise in both the magnitude and the spatial pattern of global temperature observed over the twentieth century pretty accurately. They also reproduce the changes we've actually observed in sea ice, ocean temperature, the hydrological (that is, water) cycle, and so on. This increases our confidence that the models are a reasonably good representation of the real climate, and they're taking all of the most important factors into account.

Models have also been around long enough to be tested with actual predictions. In 1988, a climate scientist at NASA used a basic model to simulate changes in temperature over the coming couple of decades. Over the next twenty years or so, the real-world climate acted very much like the simulation. And that was using a model that's much more primitive than those in use today.

Some people are skeptical about climate models because they know that weather forecasts are good for only a few days into the future. Even so, scientists think they can project climate decades ahead (with one important caveat: nobody knows what future greenhouse-gas emissions, deforestation, and other human factors driving climate change will be over the next century).

Once they make an assumption about future greenhouse emissions, however, modelers are confident that models give useful information about future climate. If that sounds impossible, it's because many people imagine that climate modeling is just an extended type of weather forecasting. But while climate models are related to weather models, they're doing different things.

To understand the difference, think about trying to predict whether a coin flip will come up heads or tails. It's impossible to do that for any single flip (which is something like a weather forecast for next month). But you can safely predict that if you repeat the coin toss a thousand times, you will get about five hundred heads and five hundred tails. Weather is about single events; climate is about long-term averages. While we know we can't predict the timing of specific weather events far in the future, we should be able to predict long-term averages (and other properties like climate variability).

Still, models aren't perfect. It's still impossible, for example, to get all the complexities of cloud behavior into the models in a realistic way. So climate scientists are always looking for new ways to refine their models and to estimate how far off they might be. They're making progress, but only slowly. In the meantime, decision makers who want to use information about future climate have to make do with imperfect information.

41

Since We Don't Know Whether and How Much People Might Cut Greenhouse-Gas Emissions, It's Hard to Know Exactly How High the Temperature Will Go by 2100.

Yogi Berra is credited with having said that "predictions are hard to make, especially about the future." It's true, but climate scientists do their best anyway. Computer models can do a pretty good job of simulating real-world events, but they can't ever replicate them exactly (you've probably noticed this firsthand if you've ever seen a computer-animated human in motion). Even well-understood phenomena, like the physics of flight, can't be simulated perfectly, but aeronautical engineers know they can rely on flight simulations because they know how closely their models represent the real world.

Similarly, models of the planet's climate can't simulate today's climate precisely, but they do a good job of approximating it (and they're getting better, as the models are continually improved and tested). The places where models and reality disagree help point out the areas of uncertainty.

These uncertainties come from the fact that the local climate is influenced by many different things, including ocean currents, ice cover, vegetation, and cloud cover, to name just a few. As temperatures rise, each of these can change, leading the other parts of the system to react, leading to still more changes. Right now, scientists can't reproduce all of these changes in their models, so they have to make approximations. For instance, cloud cover is likely to change as the temperature goes up. The changes could lead to extra warming, or to some cooling, or some of both. Scientists aren't certain at this point about what the total effect will be, because cloud processes take place on too local a scale for models to handle. The balance of observational evidence so far suggests that clouds are likely to cause some extra warming. But it's unclear how large this effect will be.

Since the models can't yet give a dead-on representation, scientists know they can't represent the future climate flawlessly either. That's why projections for warming and sea-level rise and other changes are given as ranges, not specific numbers. But even if the models were perfect, they still wouldn't be able to tell us how much the climate will change over the next century to the number. That's because the amount of climate change will depend on human actions. For the last several decades, human greenhouse-gas emissions have been the biggest human-caused force changing global climate.

The amount of climate change we can expect, therefore, depends on just how much CO_2 (and other greenhouse gases) humans will emit in the future. Since no one can predict the choices individuals, businesses, and governments will make, climate models cover their bases by running several different simulations to project rates of warming over the next century. Each simulation is based on a different scenario involving the world's population, economy, technology, and energy use. Those possible futures all translate into different amounts of emissions, so each scenario results in a different amount of climate change later in this century. Until about 2050, it doesn't matter much what we assume about future CO_2 emissions, because much of the warming until then will be a delayed result of past emissions and because the different emissions scenarios are pretty similar early in the twenty-first century.

The results of these different scenarios probably won't come as a huge surprise: the models show that if we end up cutting emissions by a lot, our climate will change significantly less than if we stick to business as usual. As shown in the latest IPCC report, from 2007, the scenario based on the lowest emissions estimate led to about 3.2°F of overall warming from 2000 to 2100. The scenario based on the highest estimate translated to 7.2°F of warming.

Note, however, that different climate models come up with different numbers, even with the same emissions scenario. So the 3.2°F and the 7.2°F represent an average of what the models say. In fact, some of the models say it will be less than 3.2°F on the low end and more than 7.2°F on the high end.

It's important to understand that climate models are constantly being improved. That's partly because computers keep getting more powerful and software keeps getting more sophisticated. It's also partly because our observations of the actual climate are improving, so modelers can test their models better. The 2007 IPCC report was based on the best available models and information at that time, but there have been plenty of improvements since then.

The point is that projections are always a snapshot of a science that keeps evolving. Today's projections are more reliable than earlier ones, but as models continue to improve, we'll get new projections that will make these obsolete.

42

An Imperfect but Still Pretty Good Prediction: Sea Level Will Rise Two to Six Feet by 2100. But That Could Change.

There's no doubt that sea level should rise if Earth's average temperature goes up. For one thing, water expands when it warms up; warmer water simply takes up more space than cooler water. For another, mountain glaciers and ice sheets (like the ones covering Greenland and Antarctica) will tend to melt more in summer than they do now because summers are likely to be warmer in the future. (Ice floating on the sea, like the ice that covers the Arctic Ocean, will also melt. But that doesn't add to sea level, any more than melting ice cubes make a drink overflow.)

How high sea level will be in 2100, though, depends on how much warming there will be. Since scientists don't know exactly how much the planet will heat up, they can't say exactly how much sea level will rise. The best they can do is give a range of likely numbers, and as of now their best projections range from about two feet at the low end to six feet at the high end, on average. This doesn't mean it couldn't be less or that it couldn't be more. It just means anything less than two or more than six feet is less likely.

There's one more factor that goes into sea-level projections besides expanding water and melting ice. Ice doesn't just melt; it also flows very slowly (at a glacial pace, you might say) from the ice sheets covering Greenland and Antarctica down to the sea, where it

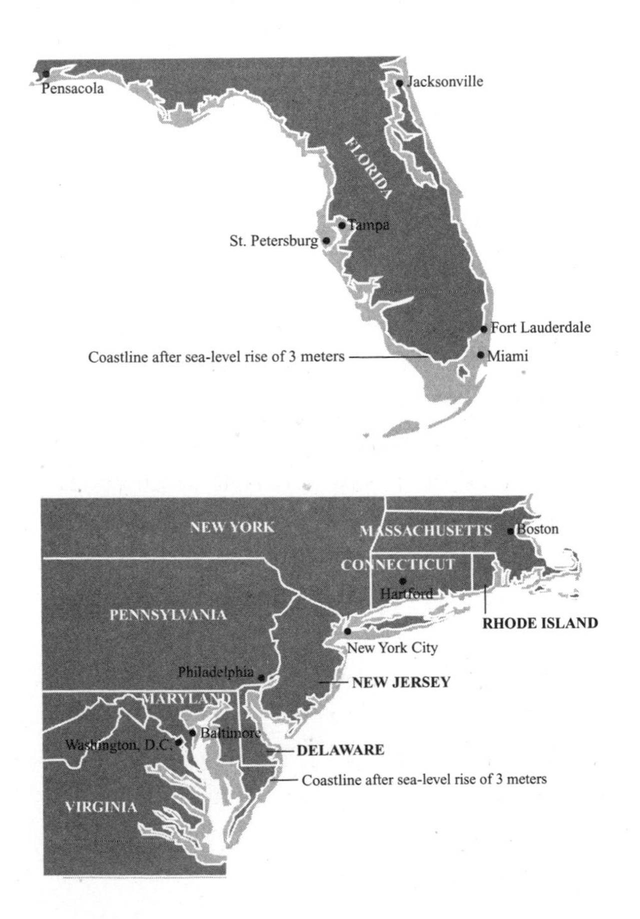

eventually breaks off into icebergs. Glaciologists (scientists who study glaciers) have known for years that as the world warms, those glaciers might move faster. Water trickling down from the surface might make the rocks far below more slippery. Or warmer ocean water might melt some of the ice down where the glacier meets the sea, making it easier for the rest of the glacier to flow.

In 2007, when the most recent IPCC report came out, the authors had no climate model simulations available for them to assess that included the latest understanding of glacier processes. The report projected sea-level rise by 2100 at between six and a half inches and two feet but noted that this projection excluded future rapid changes in ice flow.

Very soon after the IPCC report was issued, satellite observations detected some of those changes starting to happen. At the same time, scientists who create virtual-reality climate simulations figured out ways to get a better idea of how sea level and temperature are related. Taken together, these new pieces of information helped climate scientists to come up with numbers they feel more confident about. A more recent report produced in the United States, called *America's Climate Choices*, includes the more up-to-date two-to-six-foot range.

But those numbers could change in the future as glaciologists refine their understanding about how ice sheets respond to climate.

43

The Effects of Greenhouse Gases Won't Magically Stop in 2100.

When scientists show graphs of increasing temperatures and rising sea levels, or project changes in extreme weather events, or talk about disruptions to ecosystems, they usually refer to what's going to happen by 2100—the end of the current century.

That might give the impression that 2100 is some sort of magic year—that the climate will be done changing, and we'll be able to start adjusting to whatever the world looks like at that point.

That's completely wrong, of course. It's just a convenient date. People can wrap their minds around it, and, perhaps more important, it gives climate models a manageable cutoff point. If the models were called on to project conditions two hundred or three hundred or five hundred years into the future, they'd take a lot longer to produce any results, and those results would be increasingly uncertain.

But if emissions continue to climb, and CO_2 and

other gases continue to build up in the atmosphere, global temperature will continue to rise along with them. If the buildup is still going on after 2100, the amount of heat trapped by greenhouse gases will continue growing as well.

In order to make 2100 a truly magic year—that is, a year when greenhouse gases are at their greatest level in the atmosphere—emissions would not only have to stop growing but would have to diminish to almost nothing. Climate scientists sometimes compare our atmosphere to a bathtub with a very tiny drain. For thousands of years, the trickle of CO_2 entering the atmosphere from natural sources was balanced by the trickle that naturally drained out. Since the Industrial Revolution, when we started burning fossil fuels, we've opened the faucet wider and wider, so the water level (or, in the real world, the CO_2 level) has risen.

If we simply stopped opening the faucet any wider, the water would still continue to rise. If we turned it down, the water would still continue to rise, just more slowly. And if we turned it back down to the original trickle—which would mean stopping all human emissions—the water level would stop rising.

But it wouldn't fall, because that tiny drain is still tiny. Most of the extra CO_2 we've added would stay in the atmosphere for a very long time. As a result, the planet would very likely remain in a state of higher temperature, altered weather patterns, higher sea level, and the rest for at least a thousand years, and the world would probably keep changing. The ice on Greenland, for example, is already shrinking measurably. If the temperature goes higher, as it almost certainly will, and if it stays higher for hundreds of years, the ice will keep on shrinking long after 2100.

All of this imagines that we turn off the human emissions faucet completely. Since it's hard to conceive of how that could happen, the idea that 2100 will mark some sort of milestone—which scientists have never meant it to be in the first place—is clearly not correct.

It's also important to point out that temperatures keep rising even after atmospheric CO_2 stops increasing. That's because it takes a long time for the ocean to adjust to the atmospheric CO_2 concentration.

44

Best Guess About Atlantic Hurricanes in the Future: Fewer, but More Powerful.

Hurricanes draw their power from the heat in tropical ocean water. Sea surface temperatures in the part of the Atlantic where hurricanes form and intensify have been rising for the past century or so, at least partly because there's more and more heat-trapping CO_2 in the atmosphere.

Put those together, and you'd probably expect the number of hurricanes to be on the rise as well. You'd probably also expect the average hurricane today to be more powerful than it was decades ago.

But that's not necessarily so. Hurricane formation is a very complex process. You need warm oceans, but you also need favorable winds. If the wind is blowing in different directions or at very different speeds in different layers of the atmosphere, hurricanes can have a hard time getting started even if the ocean is very warm. Differences in ocean surface temperatures between different basins—the Atlantic versus the

Pacific, for example—also have an effect on hurricane formation.

So while the Atlantic is getting warmer over time, and while the number of Atlantic hurricanes has grown in the last twenty years, it's not clear that rising CO_2 levels have caused the increase. Scientists think that changes in other factors could be playing an important role. If you try to go back even further in time, using historical records, things get even harder to unravel. That's because some hurricanes are born and die without ever touching land. The only way to be sure you're counting them all is with satellites, and we've only had those available since the late 1960s.

It's also not clear what might happen over the coming century as the oceans keep warming. The best computer simulations that now exist still can't capture all of the complexity of hurricane formation, so it's unwise to count on them too much. But thus far, they tend to point to a future where we would actually see fewer hurricanes in the Atlantic.

The simulations also say, however, that even with fewer hurricanes overall, the most destructive storms—known as Category 4 and 5 hurricanes, with wind speeds over 131 miles per hour—could get more common and more intense. Again, this is still a provisional result, so we'll have to wait for better simulations and more observations before the question can be considered settled.

When a hurricane sweeps across the Atlantic and slams into the Eastern Seaboard or the Gulf Coast, as Irene did in August 2011, the amount of damage it does and the number of people it harms depend on where it strikes, not just on how powerful it is. If it comes ashore in a place where not many people live—as Hurricane Alex did when it struck about a hundred miles south of Brownsville, Texas, in June 2010—it doesn't do a lot of harm. When it makes an almost direct hit on a major city (Katrina, 2005), it can be expensive and deadly. But even if the Atlantic were to see a decrease in strong hurricanes, damage and death would likely get worse anyway because more and more people are living in harm's way.

In 1960, about fourteen million Americans lived along coastlines where hurricanes are known to strike. By 2009, that number had jumped to more than thirty-two million. More people lived in Harris County, Texas (which includes Houston), in 2005 than lived along the entire Florida coast fifty years earlier.

And so on. You can find dozens of statistics like that, not just for the United States, but for the rest of the hurricane-prone world as well, including much of the Pacific coast of Asia (where they're called typhoons) and the northern Indian Ocean. In all of these areas, population has grown dramatically over the past half century, but the growth has been greatest in major cities along the coasts—where hurricane or typhoon danger is the greatest.

If more people and property are in hurricane-prone regions, more damage will likely occur, whether or not hurricanes get more powerful. If hurricanes become more powerful, as current research suggests, so much the worse. A Category 5 hurricane has 80 percent more destructive power than a Category 4, so even if the number of hurricanes goes down, the damage can still be greater overall if the number of very intense hurricanes increases.

45

Whatever Happens with Hurricanes, Higher Sea Level Will Make the Storm Surges They Cause More Destructive.

Most people probably think the greatest danger from a hurricane lies in the storm's punishing winds and torrential rains. But those who live on hurricane-prone shores, including the U.S. coastline from Texas to North Carolina, know there's something else to worry about. It's the storm surge, a huge pulse of seawater pushed up onshore by the oncoming hurricane, like a slow-moving tsunami. Just like a tsunami, a storm surge can pose a terrible threat to people and buildings. Much of the devastation caused by Hurricane Katrina in 2005, for example, came when the surge of Gulf water overwhelmed the protective levees to flood the city.

As Earth warms up, it's possible that hurricanes could get stronger, but even if they don't, the damage they do may get worse anyway. The reason is sea-level rise caused by melting ice and the expansion of seawater as it warms. On average, the world's oceans are expected to rise by up to several feet by the end of the twenty-first century. If that happens, storm surges will have a built-in head start on any destruction they can wreak. Just imagine if sea level in the Gulf of Mexico had been several feet higher to start with when Katrina came along and pushed a wall of water ahead of it.

Of course this scenario assumes that all other factors besides sea-level rise remain the same. In reality many factors might not. For example, there's the possibility that higher average sea levels will be coupled with fewer but more intense storms. If development along coastal areas was more strictly regulated, we could actually see less devastation to human populations in urban areas by hurricanes. Of course, in areas without strict building codes or other adaptation strategies, the dangers associated with intense weather patterns are all the greater.

In fact the insurance industry has already been thinking about the problem. In a study that came out in early 2010, it was estimated that sea-level rise will make storm surges cause 20 percent more property damage by the year 2030. But that's if hurricanes don't get any stronger and if no more buildings, roadways, and other expensive construction happen along the coast. If hurricanes *do* get more powerful, or if more hotels and condos *do* go up near the beach, and no man-made or natural barriers (like barrier beaches) are in place, the damage will be even greater.

46

Climate Change Will Force People to Move, but Whether It's a Million People or a Hundred Million Is Hard to Say.

People have almost certainly been forced to relocate in response to climate change, probably for as long as the human race has existed. That's likely to continue during the coming century as rising temperatures threaten freshwater supplies, drive up sea level, and cause other disruptions to what we have come to think of as "normal" climate. In low-lying countries such as Bangladesh, rising seas could displace many thousands of people, and low-lying island nations such as the Maldives in the Indian Ocean may even become uninhabitable. In early 2011, about two thousand people were forced to abandon their homes on the Carteret Islands of Papua New Guinea as a result of sea level rise.

Putting a reliable number on how many people will be displaced by climate change by 2100, however, is very difficult to do. One reason is that it depends in part on human behavior, which is unpredictable. Some people might give up and move if they have to walk three miles to the nearest freshwater supply. Others might not.

Even in cases where climate change is a factor, there could be plenty of other things going on. Political instability, overpopulation, poverty, and war can also play important roles in driving people from their homes, and to complicate things further, it's quite possible that climate change could make political instability, poverty, and conflict more likely. One study, for example, has shown that warmer temperatures are historically associated with a higher risk of civil wars in Africa and projects as much as a 54 percent increase in the average likelihood of conflict by 2030. Some experts have also argued that the conflict that began in Darfur, Sudan, in 2003 came about in part due to climate-change-related drought. The mix of factors, however, will be different from one country to the next and even from one region to the next within a single country.

Because of all these uncertainties, predictions about how many people will be displaced by climate change range all over the place. One aid organization has pegged the number at one billion; the economist Jeffrey Sachs, who is director of the Earth Institute at Columbia University, has cited a figure of "hundreds of millions," and a recent study by the International Institute for Environment and Development suggests that relatively few "climate migrants" are likely to move any significant distance at all.

It's safe to say, therefore, that climate change will be a factor in future migration, but it's extremely difficult to say with any accuracy how many people this will affect.

47

Climate Change Can Be Bad for Your Health.

When people worry about the possible dangers of climate change, they don't usually think in terms of their personal health. More often, they're thinking about the economic damage from rising seas, or about increased drought and other extreme weather events, or about endangered animals.

But climate change is already leading to a higher risk of injuries, illnesses, and even death, from a wide range of causes. According to a 2004 report by the World Health Organization, over 150,000 extra deaths worldwide in the year 2000 can reasonably be blamed on climate change, and the vast majority of these fatalities were children. This number is likely to be low because the WHO only looked at a few of the known health effects of climate change. Many more people are suffering illnesses and injuries that will be affected, for the worse, by a changing climate. Unless individuals and public-health and health-care systems are better prepared to cope with these health risks, the climate-change-related death rate is projected to increase.

One way that climate change can directly impact health is through heat waves. Climate scientists expect to see more heat waves, and more intense heat waves, as the planet's average temperature rises. During an especially punishing series of heat waves that struck Europe in 2003—stronger than any over the previous five hundred years—about seventy thousand more people died than would have been expected, many of them elderly or very young, or people of all ages who also suffered from conditions such as cardiovascular or respiratory diseases.

It's not possible to blame any single heat wave on climate change with complete certainty. However, analyses demonstrated that human influences on climate made the 2003 European heat wave at least twice as likely as it would otherwise have been. Because heat waves are likely to come more often and last longer as we approach the end of this century, the number of heat-related deaths is likely to increase. Increasing the number and effectiveness of heat-wave early-warning systems, improving public understanding of the risks of heat waves and the protective steps that individuals need to take during a heat wave, and ensuring access to cooling shelters will decrease the projected mortality in a warmer world. Although air-conditioning is frequently mentioned as a solution, reliance on it will not be feasible in most developing countries. Further, depending on the source of electricity, air-conditioning can increase greenhouse-gas emissions, feeding climate change.

While cold waves and frigid temperatures will come

less often in a warming world, this is unlikely to significantly change the number of cold-related deaths, and any decrease will not offset the rise in deaths from heat.

Another climate-related health risk comes from air pollution—especially ozone. It is important to distinguish the ozone in the stratosphere that protects us from the sun's harmful UV rays from the ground-level ozone that burns the eyes and irritates the lungs. High concentrations of ground-level ozone can trigger asthma attacks and affect people with other chronic lung diseases. More deaths occur on days with higher ozone levels. Ozone forms most easily in hot weather, with the rate of formation higher with higher temperatures. Many American cities have "ozone alert" days, where people at risk are urged to stay indoors. With higher summer temperatures and more heat waves, ozone-related health problems are projected to increase as well. And longer summers are also expanding the growing season for plants such as ragweed, whose pollen can trigger allergic reactions, including asthma attacks.

Infectious diseases are another concern. Mosquitoes and ticks can carry diseases such as malaria, dengue fever, and encephalitis that they then transmit to people. As temperatures rise, mosquitoes and ticks can expand their geographic range, which could expose more people to the risk of these diseases. Whether or not people actually experience these diseases depends on the ability of the public-health systems to prevent the mosquitoes and ticks from acquiring the diseases in the first place and, if that fails, to control the spread of disease-carrying mosquitoes and ticks. Temperatures have always been warm enough in the southeastern United States, for example, for the mosquitoes that carry dengue and malaria. Although malaria was a problem many decades ago, it is very rarely a problem today, in large part because of improvements in standards of living. Dengue, on the other hand, has been making a comeback in the United States. The situation is different in places like Russia where the public-health system is less robust; the expanding ranges of disease-carrying insects and other animals could become a serious health problem.

Another set of illnesses, including cholera and typhoid fever, are transmitted by contaminated food and water. The increase in heavy rains and floods projected for a warming world will raise the risk of outbreaks, particularly in places with inadequate sanitation and public-health systems.

The greatest concern in developing countries is that changing patterns of rainfall, including floods and droughts, could disrupt food crops, leading to hunger and malnutrition in poorer countries. Malnutrition is associated with 50 percent of all deaths in children. Malnourished children are more susceptible to diarrheal disease, malaria, and other leading causes of death. These diseases are themselves sensitive to changing weather patterns, possibly interacting to cause even more disease.

Climate change will make it more difficult to control the numbers of cases of injuries, illnesses, and deaths due to a wide range of diseases. The extent to which increases occur will depend on other factors, particularly socioeconomic development and increased investment in public-health and health-care systems. Without additional investment, increasing numbers of children are likely to be affected by a changing climate.

48

Climate Change Can Be Bad for the Health of Entire Species, and Even for Their Survival.

Biologists have been describing and naming the new life-forms they've come across for more than three hundred years. So far, over 1.5 million species have been named, but thousands more are discovered each year. Estimates for the grand total generally range from 5 to 30 million different species, with some as high as 100 million. That's a lot of diversity, but for every species of plant or animal living today, the fossil record contains hundreds of others that have gone extinct over the course of the planet's history.

Many of those previous life-forms died out during "mass extinction events"—relatively short periods when rapid changes in Earth's atmosphere or climate led to major die-offs and a steep decline in the planet's overall biodiversity, or variety of life. There have been many mass extinctions since the dawn of life on Earth, but the term usually refers to five particularly catastrophic "Big Ones." The last Big One occurred about sixty-five million years ago and is believed to be largely due to the devastating aftereffects of an asteroid impact. This mass extinction spelled the end of over 30 percent of the planet's species, including the dinosaurs.

No one is expecting another asteroid collision anytime soon, but some ecologists and biologists think that human activities are having an asteroid-like impact on Earth's biodiversity. Scientists have documented some seven hundred species that have gone extinct over the past couple hundred years, mostly due to habitat destruction from clear-cutting of forests for agriculture and other changes to the land (though other factors like pollution and unsustainable hunting and fishing practices have also played a role).

But these recorded extinctions are almost certainly a tiny fraction of the total number of species that have been lost in the last couple hundred years. Most of the extinctions we know of have affected mammals, birds, and amphibians. We know very little about all the plant and insect species that have also likely disappeared. And although human activities continue to whittle away at the amount of available habitat, species are also under pressure from the rapidly warming temperatures and shifting precipitation patterns caused by human-induced climate change.

Some species can live comfortably in a variety of climatic conditions, but others have evolved to live within a much narrower range. As those conditions change, hundreds of species are shifting to higher latitudes or altitudes to stay within their comfort zones. Having to move isn't the end of the world—unless, of course, there's nowhere to go or no way of getting there. Species that don't generally move far or quickly, like amphibians, will have trouble getting where they need to go. Also, many populations that are hemmed in by habitat destruction and physical barriers like highways, or that already live on mountains and near the poles, will be unable to find cooler climates.

The dual pressures of climate change and habitat destruction are having a clear, irrefutable impact on populations of all kinds, but that doesn't mean it's easy to make projections about future extinctions. For one thing, no one knows exactly how much Earth will warm in the coming decades, in part because no one knows when or even whether we're going to curb our emissions of greenhouse gases. There's also uncertainty about the future of land use: Will we develop land in a way that makes it possible for species to move, or not?

Finally, scientists don't know the extent to which various species will be able to evolve and adapt to climate change. Laboratory experiments with fruit flies show that fast-reproducing, short-lived species can evolve rapidly in response to changing temperatures. For many plants and animals, though, climate change is happening much too quickly for species to keep up.

Given these unknowns, researchers have used computer models to estimate the rate of future extinctions under different warming scenarios. In a situation with minimal warming, they project that nearly 15 percent of the planet's species will be committed to extinction by 2050 (that is, they may still be around, but with essentially no chance of long-term survival). Under a maximum-warming scenario, the number is around 40 percent. The numbers could be higher or lower depending on species' ability to adapt or move.

The fact that extinction rates in recent years are already an estimated fifty times higher than typical rates recorded in the fossil record, and the fact that they're going to keep rising, have led many scientists to conclude that Earth is already in the midst of a sixth mass extinction event.

As species go extinct, we not only lose them forever but also lose their contributions to our ecosystem, such as providing fertile soils, clean water, food, and medicines. The anticancer drug tamoxifen, for example, was first discovered in the bark of the Pacific yew tree. If that tree had been driven to extinction, the treatment might never have come to light. As temperatures rise and species are lost, we will lose opportunities to find other new cures, and we will also lose many other critical functions provided by plants, animals, and microbes.

49

Freshwater Will Become Scarcer.

Water levels in the Ogallala Aquifer, a vast underground reservoir that lies beneath the Great Plains in the United States, have fallen by over two hundred feet in some parts of Kansas and Texas since around 1950. The reason is simple: demand is greater than supply. The water in the Ogallala comes from rain that fell over many thousands of years. Some of that rain soaked into the soil and gradually made its way down into small holes and cracks in the sandstone formations below the surface. Rain is still trickling down into the aquifer, but nowhere near fast enough to make up for the water we're pumping out, mostly for crop irrigation.

In many places in the world—especially in dry regions with growing populations—water shortages are already a serious problem. Rivers that used to run all the way to the sea are literally drying up before they get there, as people siphon more and more water upstream. The Yellow River in China, the Colorado River in the United States, and the Ganges River in

India all run dry for part of the year due to overuse. Even without climate change, those shortages would get more serious as the world's population continues to boost the demand for water.

But scientists expect climate change to make matters worse. One reason is that the majority of the world's glaciers are shrinking and snow cover is decreasing as more winter precipitation falls as rain rather than snow. Over one-sixth of the world's population depends on water from snowmelt, but that supply is becoming less reliable: warmer temperatures mean an earlier spring melt, which leads to an earlier surge in snow-fed streams, which means lower water flow later in the year. By the time summer comes and the need for water is greatest, the snowmelt tends to be long gone.

Besides changing the time when water from melting glaciers is available, climate change will ultimately affect how much water can be obtained from glaciers. Initially, as glaciers melt, there will be more water, but after the glaciers disappear completely, there will be none at all.

Increasing temperatures also cause more water to evaporate from lakes and rivers, and from the soil. This makes clouds form faster, which in turn causes more precipitation. In other words, climate change is causing the water cycle to speed up. That may sound like good news, but it isn't, because a faster water cycle means more flooding and more droughts. Floods are similar to the early-snowmelt problem. They provide a burst of water all at once, but it tends to run off before the land can absorb much of it. The combined result of more floods and more evaporation is that less water makes its way into the groundwater supply.

In coming decades, warmer temperatures will have the biggest impact on places that are already water stressed. One example is the western United States. California's Central Valley grows a big fraction of the nation's fruits and vegetables, but farmers have to rely on water from hundreds of miles away for irrigation. Water is pumped into the Central Valley from reservoirs fed by snowmelt in the Sierra Nevada. With less snowmelt or earlier snowmelt, even if there's no change in precipitation amounts, there is less water available in summer for irrigation. This is already happening: flows during late spring and summer have decreased on the rivers that provide much of California's water. There's no strong consensus about whether overall precipitation is expected to increase or decrease in California, but somewhat more climate models predict decreases than increases. If those models turn out to be right, that will make the situation even worse.

In the southwestern United States, the situation is more dire. Here, there is a strong consensus among climate models toward less precipitation coupled with more evaporation because of warming. The net result is a lot of water stress.

The East won't necessarily escape unscathed, either. Despite increasing precipitation in most places, some areas of the eastern United States—especially Florida—are projected to experience more water stress in future decades due to population growth, which puts pressure on underground water supplies that are already being used faster than they're getting replenished.

The good news is that in many places we use water very inefficiently, which means it's possible to cut way back without undue suffering. Agriculture and industry account for over 90 percent of the world's freshwater use, but conservation strategies such as more efficient irrigation methods and wastewater recycling have already helped some places make big cuts in water use. Despite a steadily increasing demand on the public water supply, these agricultural and industrial conservation strategies led to the United States using 5 percent less water overall in 2005 than in 1980. Switching away from relatively low-value but thirsty crops, like alfalfa in Southern California, can be another way to adjust to greater scarcity with relatively small economic impact.

50

Droughts Will Probably Come More Often.

Many people probably think of a drought as an extended period with little or no precipitation. But in some regions and seasons—for example, most of California in summer—little or no precipitation is normal. So a better definition of drought would be "an extended period with below-average precipitation." Climatologists consider that the southeastern United States, for example, had a drought in 2010 because the region received less than the average precipitation in that year. But the amount of precipitation that counts as a drought in the Southeast would count as well above normal for desert areas in the Southwest—so much above normal, in fact, that it would cause problems.

But precipitation alone is only part of the story. Even with normal or near-normal precipitation, higher-than-average temperatures can force moisture to evaporate from the soil. This can cause the same sorts of problems that below-average precipitation causes. So for many purposes, a definition of drought based on below-average soil moisture—which can cause problems for agriculture, plants and animals, and human water supplies—is the most useful. The widely used Palmer Drought Severity Index is essentially a measure of soil moisture content. The famous dust bowl of the 1930s, for example, the most severe U.S. drought on record, was caused by a combination of below-average precipitation and above-average temperatures.

Using a definition of drought based on soil moisture, projections show that both long- and short-duration droughts are expected to occur more often nearly everywhere in the United States—although this change is likely to be relatively small in Alaska, where projected increases in precipitation will partly offset the drying effects of warmer temperatures. This is an example of a region where more drought is expected even though precipitation is projected to increase. The Southwest is expected to experience both increased temperatures *and* decreased precipitation—double trouble as far as the risk of drought is concerned.

51

Climate Change Is Likely to Destabilize the Food Supply.

In 2010, the world produced more than enough food to feed the planet's 6.7 billion people, but 925 million people—almost 14 percent of the global population—went hungry anyway. That's because food isn't distributed evenly. This uneven distribution could change in theory, but it's not clear how or when that might happen. So projecting the available food supply into the future is more complicated than simply trying to match up predictions about food production and population.

Even if it were that simple, it's unclear how many people the world could feed. By 2050, Earth will very likely be home to over nine billion people—nine times as many as there were living in 1900, and twice the size of the global population in 1981. Most studies of Earth's "carrying capacity"—the maximum number of people it can support with food and other resources—indicate that there is enough land to feed nine billion, but it depends a lot on the types of diets people consume and the extent to which land is used to produce biofuels, such as ethanol, rather than food.

Climate change adds to the uncertainty. Scientists expect rising temperatures and changing precipitation patterns to affect crops around the world over the next several decades. Projections show that higher carbon dioxide levels will enhance crop growth in most places, and warmer temperatures will be good for some crops in some places. However, in most places changes in temperature and rainfall are expected to reduce crop production. For more than about 1°C of additional global warming, the negative impacts of climate change are likely to outweigh the benefits of CO_2 alone. The result would be an overall rise in food prices.

There are several ways that climate changes will challenge food production. First, as we've discussed, climate change is causing an increase in how often extreme weather events like floods, droughts, and heat waves happen. All of these can badly damage or destroy crops. Second, warmer temperatures will lead to more evaporation, which, along with rainfall changes, will contribute to more frequent and severe droughts in many dry regions and a shorter growing season in the tropics. Third, warming will speed up the rate at which crops mature, meaning the plants will be rushed through the growth stages that are critical to food production.

All of these negative impacts are likely to be worse in the tropics than in cooler, temperate regions. For this reason, farmers in the tropics will struggle to compete with their counterparts in cooler countries and are likely to lose income to competitors. Because hunger is already more common in tropical countries, particularly in rural areas, climate change is an especially difficult challenge to food security.

Farmers can adapt to climate change in several ways—by breeding new crops, for example, or expanding into new regions, or adding irrigation. But each of these strategies requires fairly large and long-term investments to be successful. Therefore, the ultimate impacts of climate change will depend not only on total greenhouse-gas emissions but also on the capacity and will to invest in agriculture.

IV

CAN WE AVOID THE RISKS OF CLIMATE CHANGE?

52

Who Says a 2°C Temperature Rise Won't Bring Really Bad Consequences? Not Scientists.

For some years now, policy makers in Europe—and more recently in the United States as well—have talked about the virtues of limiting the rise in global temperature to no more than 2°C, or 3.6°F, over what it was in the mid-nineteenth century. Given how often the 2°C figure is invoked during international climate negotiations such as the 2009 COP15 conference in Copenhagen, you might assume that this is a number scientists have agreed on.

Not true. Climate scientists do agree, almost universally, that any temperature increase is likely to be disruptive. Even the 1°C increase we've seen over the past century or so has already led to a measurable increase in droughts and heat waves in some parts of the world, to changes in ecosystems, to melting ice in Greenland and other places, and more.

As far as anyone knows, however, the 2°C threshold doesn't represent some sort of magical dividing line, where things are fine if we stay under it and awful if we go above it. For some parts of the world—small island nations, for example—dangerous disruption could happen before we reach the 2°C level. During the COP15 climate negotiations in Copenhagen, some of these countries were campaigning for a target of just 1.5°C. In other parts of the world, 2°C or even 3°C or 4°C might not be a disaster.

There's always the chance that we could reach some sort of "tipping point"—a point of no return where something really bad starts happening that will be impossible to stop. But there's no consensus about how much warming it would take to do that; scientists do agree, however, that the risk of dangerous climate disruption keeps increasing the warmer the planet gets. Still, this target has been formally adopted by the European Union as a policy goal.

Thus, a 2°C rise is simply a target to aim for—although not an easy one to hit. It wouldn't have been easy to hit when the 2°C target was first proposed. With three years of emissions growth since then, and with more CO_2 now in the atmosphere, it will be much harder today. Meeting the target will require significant changes in how we produce energy and in how industrial societies operate in general. And it's getting more unrealistic with every passing year, as emissions continue to grow and the upward trend in global temperature continues.

Some scientists are uneasy with the whole idea of choosing a specific target number, since the available science doesn't support a hard threshold like this. But many agree that choosing some target is better than having none. It's something like the medical consensus on cholesterol. Doctors advise that a person's total cholesterol shouldn't go above 200, but it's not as if you'll definitely have a heart attack if you hit 210. And there's no guarantee you'll be safe if you stay at 190. What they do know is that the risk goes up as your cholesterol goes up.

53

Using Ethanol in Your Car Can Reduce Emissions—but Not Always by a Lot.

Touch a match to alcohol and it burns. That's why flaming desserts flame: they're doused with brandy. The way you make alcohol is by fermenting and distilling the sugars in plants—which goes not just for grapes but also for sugarcane, corn, and (with varying difficulty) pretty much any sort of plant material.

What burns on a dessert plate can also burn in an automobile engine, and plant-based alcohol, usually in a form known as ethanol, is probably the most familiar example of what are known as biofuels. Ethanol already makes up 10 percent of the liquid fuel going into American cars. Burning ethanol has a number of advantages over burning gasoline. For one thing, it can be homegrown in the United States rather than imported from somewhere else, and it's renewable. You can always grow more of the raw material—that is, plants of various kinds. You can't grow more coal or oil. For another, using ethanol can reduce carbon emissions.

It's not that ethanol is emission-free, however. Like gasoline, ethanol is a carbon-based fuel, so when it burns, it does release CO_2 into the atmosphere. But that carbon originally came *from* the atmosphere. Plants breathe in CO_2 in order to grow. If plants are taking carbon out of the air and burning the ethanol made from plants puts that same carbon back into the air, you might think everything would come out more or less even—that overall, there should be no net addition of CO_2 to the atmosphere.

But things are more complicated than that. Most ethanol in the United States is made from corn kernels (in Brazil, by contrast, they use sugarcane). But growing corn takes energy—for plowing fields, for manufacturing fertilizer, for harvesting the crop. It also takes energy to ferment the corn, distill out the alcohol, and transport it to places where it can be mixed with gasoline (most cars in the United States can't run on ethanol alone). Right now, pretty much all of the energy that goes into producing corn-based ethanol and moving it around comes from fossil fuels, and that generates a lot of extra CO_2 emissions. When these extras are added in, the total emissions you get from using corn ethanol may not be much lower than the emissions from just using gasoline in cars to begin with.

That's not all. When a farmer grows corn to make ethanol, more corn needs to be produced for food somewhere else. If land currently covered in forest is cleared to plant corn, greenhouse gases are released as trees are burned or decomposed and as the soil is tilled. If instead of clearing new land, farmers can simply squeeze more corn out of the land they're already using to grow enough for both ethanol and food, these extra greenhouse-gas emissions would not occur.

But corn kernels aren't the only potential source of ethanol. The plant material left over from harvesting, like cornstalks and corncobs, can also be converted to ethanol. These parts of the plant are made mostly from cellulose, so this is called cellulosic ethanol. Other, nonfood plants can also be used to make cellulosic ethanol.

If cellulosic ethanol is made from plant material that doesn't need annual plowing, seeding, and fertilizing to grow (like perennial grasses, or waste wood from forest clearing), it takes relatively little use of fossil fuels to produce. That makes its overall cost in greenhouse-gas emissions much lower than that of corn ethanol.

Another way of making biofuels is by using bacteria or algae as natural factories. Scientists are working on genetically modifying algae (familiar to most people as pond scum) to churn out all sorts of fuels, including diesel, methane (the main component of natural gas), and jet fuel, and they're working on other fuel types as well.

One problem still facing many of these biofuels is that the technology used to manufacture them is new—in some cases, just a year or two old—and needs more research and development before these biofuels can be a substitute for gasoline or diesel. There aren't any big factories making them in an economical way. With the price of regular gasoline and diesel still less than biofuels, there isn't a lot of demand from drivers yet.

So while biofuels have the potential to reduce carbon emissions, it's just not necessarily going to happen right away.

54

Burning Coal Doesn't Necessarily Mean Emitting Greenhouse Gases.

Coal is the dirtiest of fossil fuels when it comes to carbon emissions. When it's burned, coal emits more carbon dioxide than either oil or natural gas. But there's a way to use coal without generating much CO_2 at all. The process is sometimes called "clean coal"; a more technical term is "carbon capture and storage," or CCS.

The basic idea is very simple: as coal burns, the exhaust can be filtered in one of several possible ways to strip out the carbon dioxide before it goes out the smokestack. The CO_2 is compressed and liquefied and then pumped deep underground. Assuming the CO_2 stays underground and doesn't leak out into the atmosphere, CCS can transform coal from the most carbon polluting of fossil fuels to the least.

Since coal is plentiful and cheap, and because burning it to make electricity is something energy companies already know how to do, CCS has gotten a lot of attention recently. Coal-burning power plants and factories produce a significant fraction of all man-made CO_2 emissions. If they could all be converted to CCS, trillions of tons of CO_2 would be kept out of the atmosphere every year.

But CCS isn't quite as straightforward as it sounds. For one thing, filtering the CO_2 out of coal exhaust and liquefying it takes a lot of energy. As a result, you get less electricity out of every ton of coal, which makes coal more expensive to burn. The equipment for capturing and liquefying the CO_2 is also expensive to install; a coal plant using CCS costs more to build than a conventional coal plant.

Another complication is that you can't bury liquid CO_2 just anywhere. To keep it from leaking back out, it has to be pumped many hundreds, or even thousands, of feet down into specific kinds of geologic formations. These formations exist only in certain places. Another idea is to pump the CO_2 down to the bottom of the ocean, where it can dissolve into the water. Either way, large, long-distance

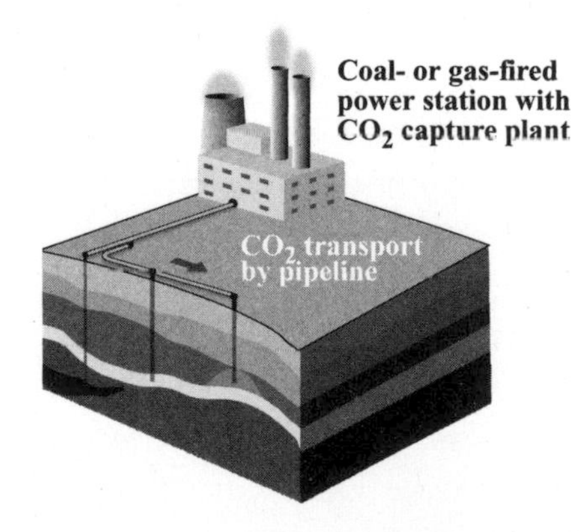

pipelines would be needed or power plants would have to be relocated to where the subsurface geology is right.

Finally, CCS is still considered an experimental technology. All of the individual parts—capturing, pipelines, and underground injection—have been tested on a small scale. But only a handful of power plants and other facilities are testing the whole process. If they can prove that the technology works as well as its proponents hope, and especially if laws someday make it more expensive for businesses to emit CO_2, CCS could become an important part of reducing carbon emissions.

55

Wind Energy Can't Solve Our Emissions Problem by Itself. Neither Can Other Renewables.

In order to reduce the amount of greenhouse gases in the atmosphere—or even just to stabilize it at today's level—the fossil fuels that produce most of it, in the form of carbon dioxide, would need to be replaced with some other, less polluting energy sources. There's no shortage of options that generate little or no CO_2 at all (such as solar, wind, or hydroelectric power). Others—specifically plant material used for energy ("bioenergy")—only put back the CO_2 the plants absorbed while they were growing. All of these energy sources have the extra advantage of being renewable. We'll never run out of sunshine, wind, rain, or the ability to grow new plants.

The problem is that coal, oil, and gas generate so much energy for electricity, transportation, and heating that it would be difficult to replace them entirely with any one of these options alone, at least not for a long, long time. The reasons have to do with technology and economics.

Here are some of the advantages and disadvantages of the most familiar types of renewable energy:

SOLAR

The sun is probably the most obvious place to look for clean power. After all, the sun beams more energy down to Earth in one hour than our planet's seven billion people use in fossil fuels in an entire year. But just because there's an excess of solar energy doesn't mean humans can rely on it in place of fossil fuels. One key issue is that all of this energy is spread out over the entire surface of Earth, so you need to devote large areas to collecting the energy.

Another reason is that while the solar panels that capture sunlight and convert it into electricity (photovoltaic panels) have been commercially available for nearly thirty years, they're still fairly inefficient. Most can convert only between 8 and 12 percent of the light that falls on them into electricity, and they are expensive to manufacture. This makes solar energy far more expensive than fossil-fuel energy. Researchers often quote a rule of thumb that the panels will need to be nearly twice as efficient (or to put it another way, the electricity they produce will have to be half as expensive) before they can be more widely used. Even then, they'll be more expensive than power from fossil fuels (unless the carbon emitted in using fossil fuels carries a financial penalty).

Another problem is that people need electricity even when the sun doesn't shine—at night or on cloudy days. Scientists haven't yet found affordable ways to store the energy from sunny days so it can be used at any time.

Energy from the sun can also be used to turn water into steam to make electricity by using mirrors to concentrate sunlight (this is sometimes called solar thermal power). This is less expensive than converting sunlight directly into electricity, the way solar (or photovoltaic) panels do, but the approach is limited by scale; it takes a lot of land to collect enough heat to produce steam.

WIND

Another abundant natural source of renewable energy is wind. Today, wind provides only about 1 percent of the planet's energy, but because we already know how to capture it with giant wind turbines, just building more turbines sounds like an obvious solution. Yet energy demand is already so high that in order to satisfy it, we would need to build hundreds of thousands of new turbines each year. Beyond that, the windiest places are often far from where people live. The cost of building all these turbines and the new power lines that would have to go with them—along with the fact that transmission lines lose more energy the farther they stretch—makes wind power more expensive than using fossil fuels today.

Wind is also highly variable, like solar energy, and unpredictable. Since the wind doesn't always blow at the times we want electricity, you need a backup power source, adding even more cost.

Finally, some people would rather not have wind

turbines nearby because they're noisy, or spoil the view, or might kill birds. At least one major wind power project off the shore of Cape Cod has been delayed for years by unhappy local residents.

At best, therefore, wind will be only a partial replacement for fossil fuels. The U.S. government projects that 20 percent of the country's electricity could come from wind by 2030.

HYDROELECTRIC

Hydroelectric dams have been used to make electricity for more than 150 years. Today, about 16 percent of the world's electricity comes from hydroelectric sources. Some countries, like Norway, Brazil, and Canada, produce the majority of their electricity this way.

But hydroelectric power is unlikely to grow much, because many of the planet's largest rivers have already been dammed, often at multiple locations. Even if all the remaining rivers were dammed, the electricity they generated would not be enough to meet the world's demand.

Moreover, hydroelectric power is not perfect. Even though a hydroelectric dam doesn't generate carbon dioxide, it can still have a significant impact on the environment. Ecosystems above and below dams are often disrupted because fish populations cannot easily travel upstream past most large dams. Dams often create huge lakes behind them, which can and do drown a lot of land. And they can trap silt that once flowed downriver, starving river deltas downstream of replenishment.

GEOTHERMAL

Geothermal power taps into the natural heat that wells up from Earth's hot interior—the same heat that powers volcanoes. Volcanoes themselves are far too dangerous to harness, but the hot rock that lies two or three miles underground in places like California and Hawaii can be tapped for its energy. Engineers force water down into drill holes, and it comes up hot enough to be used to generate electricity.

There's more heat down there than humans could ever use, but it's so deep that it's difficult and costly to get to. Only in a handful of locations can it be done at all. Extracting enough of it to substitute for all fossil fuels will be difficult, if it's even possible.

Finally, the term "geothermal energy" also, and confusingly, refers to an entirely different technology that's becoming increasingly common for homes and businesses. In this case, water pipes are buried a few yards underground, where the temperature is cooler than the air temperature in summer and warmer than the air in winter. Water circulating through the pipes is therefore pre-cooled for air-conditioning in summer and pre-warmed for circulation through radiators in winter. The result is some savings in fossil fuels.

BIOMASS AND BIOFUELS

Biomass usually refers to plant material that's burned directly for energy; biofuel usually means a liquid fuel made from plant material, like ethanol or biodiesel.

The advantages in both cases are that they're renewable (you can grow more plants), they're homegrown (so we might be able to reduce some imported energy), and they don't add any extra CO_2 to the atmosphere (if they are produced without using any fossil fuels in the process).

But biomass also is not as dense with energy as fossil fuels: a pound of biomass contains less energy than a pound of fossil fuel, so you need a much greater volume of biomass to generate the same amount of power. In order to completely replace fossil fuels, farmers around the world would need to use a lot more land to grow plants for energy. They might have to clear more land, which would cause more environmental problems.

So while these forms of energy are abundant and can help lower CO_2 emissions, no one type we have mentioned can replace fossil fuels entirely. Taken together, however, and using each one where it's most practical may make it possible to cut down dramatically on how much fossil fuel we burn.

56

Energy Costs Are Likely to Rise in the Short Term if We Limit Carbon Emissions.

Most of the world gets most of its energy from fossil fuels, for two related reasons. First, fossil fuels are historically more plentiful and less expensive than other forms of energy. Second, because they've been less expensive and because we've been using them the longest, our transportation, electricity, and heating technologies are all designed largely around fossil fuels. If coal didn't exist, we might have been getting electricity from little personal windmills for all these years. If oil didn't exist, cars might run on wood pellets (a few wood-burning cars, for the record, actually exist).

There's one more reason we use fossil fuels: they pack more energy per gallon or ton or cubic foot than most other fuels. A car would need to haul along a huge volume of pellets to be able to drive as far on wood as on a tank of gasoline.

For all these reasons, it is difficult for alternative

sources like wind, solar, and biofuels of various kinds to compete with fossil fuels today. If people were happy to pay more for these low-emitting forms of energy, it would be relatively simple to switch, although it would take some time to ramp up production of solar cells, wind turbines, and the like.

In fact, some people *are* happy to pay more. That's why electricity companies often give their customers the option to choose a form of energy that's "greener" than coal or natural gas. A customer who chooses, say, wind energy isn't actually getting different electricity from a customer who chooses a cheaper option (all the electricity is coming from the grid), but if the consumer is willing to pay, the electric company will buy more of that energy from its suppliers.

But while some people are willing to pay a little more, most folks, and most businesses (especially those that use a lot of energy), are reluctant to sign up for a higher utility bill each month. Nobody can wave a magic wand and make alternative energy cheaper overnight. So if Americans decide we have to limit CO_2 emissions, fossil fuels would need to become more expensive to discourage their use. That's the idea behind proposed government policies like carbon taxes or cap and trade. Those programs wouldn't make it illegal to pollute; they would just make it more expensive, which would affect the choices that people and businesses make.

Big-time CO_2 emitters like coal plants could choose to continue emitting CO_2 and pay the price, or they could invest in new technologies to capture and store the carbon, or they could shut down and be replaced by a more expensive source like wind or solar. Whichever way they go, the cost of making the electricity increases.

The bottom line for most people—at least everyone who currently opts for the cheapest electricity possible—would be a bigger monthly bill from the utility company. This would create an incentive to use less electricity, through conservation and more efficient appliances. For drivers, it would mean higher gasoline bills—an incentive to reduce their fuel bills by driving less or switching to a hybrid or electric car, which currently costs more to buy than a comparable gasoline-powered car but uses less fuel per mile.

In the long run, if people end up buying hybrid or electric cars by the millions, the way they do conventional cars, and if utilities switch over on a large scale to wind turbines and solar cells to generate electricity, the cost of these newer technologies is likely to come down. And since the sun and the wind are free, unlike coal or oil, the higher initial cost of buying these devices will be made up, at least partly, by savings on fuel.

Moreover, other technologies and changes in behavior that can reduce energy use—adding insulation, using more efficient appliances and vehicles, changing driving habits, and many more—could more than make up for an increase in energy prices. So while the cost of energy itself would be higher, at least at first, the cost of "energy services"—the technical term for what energy actually does for us—might well end up being lower overall.

57

Nuclear Energy Is Essentially Carbon-Free. That Doesn't Mean It's Without Issues.

Nuclear power plants work by splitting radioactive elements such as uranium and plutonium. That releases energy, which is used to turn water into steam. The steam is then forced through turbines to generate electricity. Carbon doesn't enter the picture at all, which makes nuclear power largely carbon-free ("largely," because nuclear plants are massive installations constructed with a lot of steel and concrete, and it takes plenty of energy from fossil fuels to build them).

As a result, nuclear power is often cited as a way to satisfy our hunger for electricity without adding carbon to the atmosphere. It's a perfectly valid scientific claim.

Nuclear power does have some drawbacks. For one thing, it's very expensive to build even a single nuclear power plant, and it would take hundreds of new plants in the United States to make a dent in our carbon emissions. The cost has partly to do with how big and complex nuclear plants are and partly to do with all of the permits and safety regulations plant builders have to satisfy.

Another problem is that the fuel used in nuclear plants is still dangerously radioactive after it can no longer generate power. It has to be disposed of in a way that keeps the radioactivity from escaping into the environment, and because the spent fuel stays dangerous for thousands of years, keeping it safe is not a simple job. The U.S. government has been trying to create a secure disposal site for decades and still hasn't managed to do so. For now, power plants are keeping their high-level waste right on site, stored in concrete casks, but that can't go on forever. Building a lot more nuclear plants would make the waste-disposal problem that much more difficult.

Some people also worry about the safety of the plants themselves. When, in the 1980s, the plant exploded and burned at Chernobyl, in the Soviet Union, a significant amount of radioactive material escaped. But the Chernobyl plant wasn't built safely in the first place and never could have passed inspection in a developed nation. In the 1970s the Three Mile Island nuclear plant in Pennsylvania suffered a near meltdown that, while serious, released only a relatively tiny amount of radioactivity. Since then, safety and training regulations have been made a lot stronger. Most experts are convinced that whatever the other drawbacks, safe operation is not a major barrier to the use of nuclear power plants. The earthquake and tsunami that devastated northwest Japan in March 2011, however, prove that natural disasters still pose a risk to some nuclear plants.

Finally, there's some concern that a major expansion in nuclear power could lead to an increase in the danger from nuclear bombs. In order to be used for power, uranium has to be processed (the technical term is "enriched"). The uranium used in bombs requires even greater processing. But a dramatic rise in the amount of partially enriched uranium being transported around the world raises the risk that some could be diverted, or even stolen—and ultimately be used to make nuclear weapons.

Spent fuel from power plants can also be processed to extract another radioactive element, plutonium, which can be used like uranium to fuel a nuclear power plant. This sort of recycling, called reprocessing, would make nuclear power much more efficient—it's (roughly) as though you could capture your car's exhaust and extract gasoline from it to run the car farther. But plutonium is also an ideal ingredient for nuclear weapons. For that reason, the United States banned reprocessing in the 1970s. A widespread, worldwide expansion of nuclear power, however, could put more plutonium, not just more enriched uranium, into circulation.

58

Even If We Can't Reduce Emissions, Futuristic Technology Could Save Us. Maybe. And It Could Be Risky.

Reducing greenhouse-gas emissions is the most straightforward way to help slow climate change because it gets to the root of the problem—the increasing concentrations of heat-trapping gases. But if economic and political considerations keep us from cutting CO_2 emissions significantly, some researchers have proposed other ways to keep Earth from warming.

These ideas, collectively called geoengineering, involve the use of technology, either to remove CO_2 from the atmosphere or to increase the amount of sunlight reflected away from the planet. Several types of geoengineering have already been shown to work in the lab, or on a very small scale out in the natural world. To make a real dent in rising temperatures, however, one or more of them would need to be adopted on a vastly larger scale.

One way to reduce the amount of CO_2 in the

atmosphere involves dumping massive amounts of iron into the oceans to spur a population boom in algae. Since algae are plants that consume carbon dioxide, that could start pulling excess CO_2 out of the atmosphere. Eventually, their remains—or the remains of animals that ate the algae—would float to the bottom of the sea and be buried or incorporated into carbonate rocks. Scientists have tried it in open waters with mixed results, and it's unclear whether it would work on a big enough scale to make a difference. Even if it did, the amount of iron involved could prove toxic to other marine life.

In the category of reflecting sunlight, geoengineering researchers suggest taking a hint from volcanoes. Some volcanoes erupt explosively (Krakatoa and Mount Tambora, both in Indonesia, are two examples from the past few hundred years; Mount Pinatubo, which erupted in 1991 in the Philippines, was more recent though less dramatic). When they do, they can loft tiny particles of ash into the stratosphere, along with gases that form particles called aerosols. In both cases, the particles can reflect some sunlight back into space, which keeps it from warming Earth. The eruption of Mount Tambora in 1815 likely contributed to the "Year Without a Summer" that chilled Europe in 1816. Generally, the effects of a volcanic eruption on temperature last from a few months to a year or two.

Man-made aerosols could do the same thing, and some scientists have proposed sending them into the stratosphere on purpose. To counteract the warming caused by CO_2, however, these aerosols would need to be constantly replenished, since they would gradually be washed out of the sky by rain. That could pose some danger: the kinds of aerosols under discussion can also cause acid rain, which can harm plants and animals.

Another way to reflect more sunlight would be to make clouds whiter. It might actually be possible to do this by sending fine sprays of salty seawater up into the sky. The salty spray would have the effect of making water droplets in the clouds smaller than normal, making them brighter and more reflective. It's not clear, though, what this sort of cloud seeding might do to weather patterns and ocean currents, so more research will be needed before large-scale testing happens.

Yet another scheme would involve launching flotillas of mirrors into orbit around Earth—the same general idea as cloud whitening or imitating volcanoes, but with rocketry involved.

Every kind of geoengineering clearly has some potential risks, and these risks increase as the scale of the project gets larger. Scientists can foresee some of them, but there's no guarantee that they'll be able to foresee them all. Brand-new, large-scale technologies sometimes have unexpected consequences—what you might call "unknown unknowns." Besides, scientists may never know how good a job geoengineering can do at keeping the planet cool until they try it on a very large scale, and at that point it could be too late to avoid any risks.

Even if geoengineering could keep temperatures from rising, moreover, the oceans would continue absorbing some of the excess CO_2 from human emissions. That would drive up the acidity of seawater, with potentially serious harm to ocean-dwelling plants and animals.

For all these reasons, geoengineering is not considered the first line of defense against climate change.

59

If We Made It Easier for Plants and Animals to Relocate, We Might Prevent Some Species from Going Extinct.

Because of the dual pressures of climate change and habitat loss, scientists estimate that somewhere between 15 and 40 percent of the planet's species are headed for extinction by 2050 (and some projections go even higher). The good news is that there are things we can do to minimize the loss in biodiversity over future decades, even if warming due to increasing greenhouse gases goes unchecked.

Fossil evidence and biological observations have revealed that populations of plants, insects, and animals can shift to different elevations, or move north or south, to follow changing temperatures as the climate warms and cools. But over the past couple hundred years, humans have altered much of the natural landscape, turning it into a virtual minefield for would-be climate migrants. To get from one livable location to another, many populations that could once have moved easily would now have to cross large areas of uninhabitable farmland or urban land or dangerous barriers like highways or cities.

One solution proposed by conservationists is to reconnect those patches of habitat. The idea is to link protected lands so that plants and animals can travel safely to more hospitable places. Creating these links can cost a lot of money and require complex management and coordination by farmers and other landowners. But biologists and ecologists estimate that those dispersal routes could prevent the extinction of thousands of species. It's not just a theoretical idea: such corridors are already being established in many parts of the world. In the United States and Canada, for example, conservationists are working to create a safe corridor all the way from Yellowstone National Park to the Yukon Territory, more than a thousand miles away.

While such habitat networks could potentially help reduce the loss of biodiversity in coming decades, they can't save populations that are poor dispersers, too isolated from other potential habitats, or already under serious pressure from climate change. The Florida torreya, a coniferous tree found only in one tiny ravine in Florida, is already a victim of all of these circumstances. Like other land plants, this tree can shift its range only as far as its seeds are dispersed—maybe two or three hundred yards—from one generation to the next. But scientists estimate that climate zones will shift toward higher latitudes at an average of almost six hundred yards (one-third of a mile) per year over the next century.

The Florida torreya is unlikely to keep up. Even if it could, the tree doesn't just need favorable temperatures. It also needs specific conditions for survival, including plenty of shade and moist soil—conditions that don't exist anywhere close by its current habitat. To try to save the Florida torreya, a group of activists has been transplanting seedlings to new locations in places like the mountains of North Carolina. This idea, called assisted migration, is gaining ground with many ecologists and conservation biologists. But critics argue that it would be too costly and too risky to do for large numbers of species, especially since no one can be totally sure whether, or under what conditions, it will work and what kinds of negative consequences it might have on ecosystems that receive these climate migrants.

Of course, while both methods of helping species migrate could soften the damage of climate change to biodiversity in coming decades, the most effective solution would also include trying to limit emissions of planet-warming greenhouse gases and slow the destruction of habitat from activities like the clear-cutting of forests.

60

Reducing Emissions Has Benefits and Costs. But It's Hard to Pin Down Exactly What They Are.

If the roof of a house is starting to wear out, the homeowner has two choices: spend money now to replace it or be prepared to spend even more money down the road when a major leak ruins your possessions.

Deciding how to deal with climate change involves similar kinds of calculations. Investing in low-carbon energy-efficient technology such as LED lightbulbs, renewable energy, and hybrid cars will cost money now. But if we don't limit greenhouse-gas emissions, the world is likely to see higher sea level, an increase in extreme weather events, and disruptions to agriculture—all of which would prove even more costly toward the end of the century.

Unfortunately, it's not easy to calculate exactly how much those future costs will be. Computer models that try to project future climate are very complex, but they can't capture all of the details of Earth's climate system, so they can't precisely forecast how much damage rising sea level, altered weather patterns, and disruptions to agriculture will cause.

Furthermore, computer models of the world economy are also very complex and incomplete. So even if climate models could accurately forecast, for example, that precisely 30 percent of Miami will be underwater on a particular date (which they can't), it's very difficult to project how much it would cost, decades from now, to deal with such a disaster.

That being the case, combining climate models and economic models to project the cost of climate change is prone to uncertainty on top of uncertainty.

But this hasn't stopped scientists and economists from trying. They try to estimate the "social cost" of each ton of carbon emitted into the atmosphere—that is, how much, in dollars, that ton will damage society in the future. If governments decide to limit emissions, the projected cost of carbon can be used to set tax rates and other penalties to discourage emissions. It's somewhat similar to the way insurance companies estimate the likelihood and costs of future disasters in order to figure out how much to charge their customers in premiums today.

The estimates of carbon's future cost need to include not only the physical damage from climate change but also how inflation and other economic factors will change the value of money. If the homeowner with the bad roof chooses not to fix the roof, but instead invest the same amount of money, it's possible that he or she could earn enough profit to replace all the damaged items when the roof eventually does fail.

It all depends on how much the investment grows in the meantime, on how much the price of roofing materials and labor has gone up, and on what rooms in the house get inundated.

Since none of that can really be known in advance, estimates will naturally vary, and that's the case for pricing carbon dioxide as well. In late 2010, a federal commission came up with three possible costs—$34, $21, and $5 per ton of carbon—based on a worst-case, medium-case, and best-case scenario. But others, including the U.S. Global Change Research Program, have pegged the cost far higher. A British study put the medium-case figure at $85, and the German Environment Ministry says $95.

Projecting that climate change will cause damage, including economic damage, is easy. Projecting what the damage will be in economic terms—and figuring out how to use those estimates for mitigation now—is much more difficult.

Epilogue: The IPCC Is What, Exactly?

During the 1980s, scientists were becoming more and more aware that human-generated greenhouse gases might be raising Earth's temperature and otherwise changing the climate, which could lead to all sorts of problems. The United Nations Environment Programme and the World Meteorological Organization realized that if this were true, it was important to stay on top of the situation.

So in 1988 they joined forces to create the Intergovernmental Panel on Climate Change (more commonly known as the IPCC) to "provide the world with a clear scientific view on the current state of knowledge in climate change and its potential environmental and socio-economic impacts." In plain language, they aimed to gather and summarize the best available information about how climate change happens, how it is affecting and will continue to affect society, and what the options for limiting the human impact on the climate system may be.

The word "gather" is key. The IPCC itself doesn't make any observations or do any research. Its job is to collect, weigh, and summarize the work of scientists around the world who study and publish papers about these issues.

The IPCC is best known for its major studies, known as assessment reports. These are authoritative summaries of what is known—and, equally important, what is not known—about climate change. There have been four of these so far, issued in 1990, 1995, 2001, and 2007; the next one will be published in stages, in 2013 and 2014. (The IPCC also produces special reports and technical papers on topics of particular interest, but the public rarely hears much about these.)

Each report so far has been made up of three volumes, produced by three different "Working Groups," each of which covers a different aspect of climate change. Using the last report as an example: the first volume, from Working Group 1, discusses observations of past climate change, factors affecting climate change (such as greenhouse-gas emissions and deforestation), evaluation of climate models that simulate past, current, and future climate, and these projections of future climate change themselves. This volume also evaluates and summarizes research about whether humans have already caused detectable climate change. The second volume, by Working Group 2, discusses how climate change affects humans and ecosystems through, for example, sea-level rise, extreme heat events, effects on water supplies, changes in crop yields, and so on. The third major volume, by Working Group 3, discusses ways to reduce future climate change by developing carbon-free energy sources and energy efficiency.

Because these volumes are so massive (each

approached a thousand pages in the last assessment), the IPCC also publishes two summaries of each one: a technical summary, which itself can be rather long, and a shorter summary for policy makers. It also summarizes the most important messages from all working groups' reports in a brief synthesis report.

Assessment reports are written by large teams of authors who have to stick to a schedule set years in advance. These authors are not paid by the IPCC. (The IPCC has only about ten permanent paid staff members.) After being drafted, each chapter in the reports is subjected to two levels of review, to ensure accuracy, completeness, and balance. In the first phase, typically hundreds of scientists go over each chapter. In the next phase, governments and experts re-review everything.

Review editors are responsible for making sure that all review comments have been considered. This elaborate writing and review process aims at ensuring accuracy, inclusiveness, and robustness, but it also takes years, so that when the volumes are finally published, there may already be important new work that has not been included. It's also vital to understand that scientists are independent thinkers. They often disagree with each other. So it's not at all uncommon for commenters to criticize specific statements that appear in the chapters. Review editors are required to consider all comments, but not necessarily to accept them all. Their job is to look at all the research, look at where it's pointing, and summarize it.

Finally, it's important to note that these assessment reports end up with amazingly few mistakes. There are always some, since this is a human enterprise. In the 2007 report, for example, there was a mistake about how quickly glaciers in the Himalayas might melt (the number published in the report was much sooner than the science actually suggested), and another about potential flooding in the Netherlands (the Dutch government gave the IPCC incorrect numbers, which it didn't double-check). But given the tens of thousands of words and hundreds upon hundreds of studies summarized in the reports, these are very rare exceptions.

References

GENERAL REFERENCES ON CLIMATE CHANGE

Intergovernmental Panel on Climate Change. 2007. *IPCC Fourth Assessment Report: Climate Change 2007 (AR4):*

Working Group I Report: The Physical Science Basis. http://www.ipcc.ch/publications_and_data/ar4/wg1/en/contents.html.

Working Group II Report: Impacts, Adaptation, and Vulnerability. http://www.ipcc.ch/publications_and_data/ar4/wg2/en/contents.html.

Working Group III Report: Mitigation of Climate Change. http://www.ipcc.ch/publications_and_data/ar4/wg3/en/contents.html.

United States Global Climate Change Sciences Program. 2009. *Global Climate Change Impacts in the United States.* Ed. T. R. Karl, J. M. Melillo, and T. C. Peterson. New York: Cambridge University Press.

http://globalchange.gov/what-we-do/assessment/previous-assessments/global-climate-change-impacts-in-the-us-2009.

CLIMATE CHANGE BASICS

Board on Atmospheric Sciences and Climate. 2006. *Surface Temperature Reconstructions for the Last 2,000 Years.* Washington, D.C.:

National Academies Press. Accessed January 6, 2011. http://books.nap.edu/openbook.php?record_id=11676&page=2.

Climate Central. 2010. "Accelerated Ice Loss from Greenland." Accessed December 1, 2010. http://www.climatecentral.org/gallery/graphics/accelerated_ice_loss/.

Commonwealth Scientific and Industrial Research Organisation. 2009. *Historical Sea Level Changes.* Accessed December 1, 2010. http://www.cmar.csiro.au/sealevel/sl_hist_intro.html#intro.

Environmental Protection Agency. 2007. *Inventory of U.S. Greenhouse Gas Emissions and Sinks, 1990–2005.* http://www.wrapair.org/ClimateChange/GHGProtocol/docs/2007-04_US_GHG_EI_1990-2005.pdf.

———. 2011. *Atmospheric Concentrations of Greenhouse Gases in Geologic Time and in Recent Years.* Accessed November 1, 2011. http://www.epa.gov/climatechange/science/recentac_majorghg.html#fig1.

———. 2011. *Greenhouse Gas Emissions.* Accessed November 21, 2010. http://epa.gov/climatechange/emissions/index.html.

———. *High GWP Gases.* http://www.epa.gov/highgwp1/.

———. *Methane.* http://www.epa.gov/methane/scientific.html.

———. *Methane: Sources and Emissions.* http://www.epa.gov/methane/sources.html#natural.

Hansen, J., M. Sato, R. Ruedy, P. Kharecha, A. Lacis, R. L. Miller, L. Nazarenko, K. Lo, G. A. Schmidt, G. Russell, I. Aleinov, S. Bauer, E. Baum, B. Cairns, V. Canuto, M. Chandler, Y. Cheng, A. Cohen, A. Del Genio, G. Faluvegi, E. Fleming, A. Friend, T. Hall, C. Jackman, J. Jonas, M. Kelley, N. Y. Kiang, D. Koch, G. Labow, J. Lerner, S. Menon, T. Novakov, V. Oinas, J. Perlwitz, D. Rind, A. Romanou, R. Schmunk, D. Shindell, P. Stone, S. Sun, D. Streets, N. Tausnev, D. Thresher, N. Unger, M. Yao, and S. Zhang. 2006. "Climate Simulations for 1880–2003 with GISS ModelE." *Climate Dynamics* 29 (7–8): 661–96. doi:10.1007/s00382-007-0255-8.

Hegerl, G., F. Zwiers, and C. Tebaldi. 2011. "Patterns of Change: Whose Fingerprint Is Seen in Global Warming?" *Environmental Research Letters* 6 (4), article n. 044025.

Held, I. M., and B. J. Soden. 2000. "Water Vapor Feedback and

Global Warming." *Annual Review of Energy and the Environment* 25:441–75. http://www.annualreviews.org/doi/abs/10.1146/annurev.energy.25.1.441.

Hodell, D. A., J. A. Curtis, and M. Brenner. 1995. "Possible Role of Climate in the Collapse of Classic Maya Civilization." *Nature* 375 (6530): 391–94. doi:10.1038/375391a0.

Intergovernmental Panel on Climate Change. 2001. "Factors Contributing to Sea Level Change." In *IPCC Third Assessment Report: Climate Change*. http://www.grida.no/publications/other/ipcc_tar/?src=/climate/ipcc_tar/wg1/411.htm.

———. 2007. "Changes in Extreme Events." In *Working Group I Report: The Physical Science Basis*. http://www.ipcc.ch/publications_and_data/ar4/wg1/en/ch3s3-8.html.

———. 2007. "Couplings Between Changes in the Climate System and Biogeochemistry." In *Working Group I Report: The Physical Science Basis*. http://www.ipcc.ch/publications_and_data/ar4/wg1/en/faq-7-1.html.

———. 2007. "Observations: Oceanic Climate Change and Sea Level." In *Working Group I Report: The Physical Science Basis*. http://www.ipcc.ch/publications_and_data/ar4/wg1/en/ch5.html.

———. 2007. "Observations: Surface and Atmospheric Climate Change." In *Working Group I Report: The Physical Science Basis*. http://www.ipcc.ch/publications_and_data/ar4/wg1/en/ch3.html.

———. 2007. "Palaeoclimate." In *Working Group I Report: The Physical Science Basis*. http://www.ipcc.ch/publications_and_data/ar4/wg1/en/ch6.html.

———. 2007. "Projected Changes in Emissions, Concentrations, and Radiative Forcing." In *Working Group I Report: The Physical Science Basis*. http://www.ipcc.ch/publications_and_data/ar4/wg1/en/ch10s10-2.html.

Jensen, M. 2010. "Cave Reveals Southwest's Abrupt Climate Swings During Ice Age." *University of Arizona News*, January 21. Accessed December 1, 2010. http://uanews.org/node/29591.

Kerr, R. 2001. "Rising Global Temperature, Rising Uncertainty." *Science* 292 (5515): 192–94. doi:10.1126/science.292.5515.192.

Kunkel, K. E., D. Easterling, K. Hubbard, and K. Redmond. 2004. "Temporal Variations in Frost-Free Season in the United States, 1895–2000." *Geophysical Research Letters* 31, L03201. doi:10.1029/2003GL018624.

Kwok, R., and D. A. Rothrock. 2009. "Decline in Arctic Sea Ice Thickness from Submarine and ICESat Records: 1958–2008." *Geophysical Research Letters* 36, L15501. doi:10.1029/2009GL039035.

Lindsey, R. 2010. "Are the Ozone Hole and Global Warming Related?" NASA Earth Observatory Blog, Setember 14. Accessed December 1, 2010. http://earthobservatory.nasa.gov/blogs/climateqa/are-the-ozone-hole-and-global-warming-related.

Meehl, G. A.; C. Tebaldi, G. Walton, D. Easterling, and L. McDaniel. 2009. "Relative Increase of Record High Maximum Temperatures Compared to Record Low Minimum Temperatures in the U.S." *Geophysical Research Letters* 36, L23701. doi:10.1029/2009GL040736.

NASA Earth Observatory. "Paleoclimatology." Accessed December 1, 2010. http://earthobservatory.nasa.gov/Features/Paleoclimatology_CloseUp/paleoclimatology_closeup_2.php.

National Oceanic and Atmospheric Administration. 2002. "Education and Outreach: Introduction to Paleoclimatology." Accessed December 1, 2010. http://www.ncdc.noaa.gov/paleo/primer_proxy.html.

———. 2010. *Greenhouse Gases: Frequently Asked Questions.* Accessed November 21, 2010. http://www.ncdc.noaa.gov/oa/climate/gases.html.

National Snow and Ice Data Center. "Sea Ice Index." Accessed December 10, 2010. http://nsidc.org/data/seaice_index/.

Pfeffer, W. T., J. T. Harper, and S. O'Neel. 2008. "Kinematic Constraints on Glacier Contributions to 21st-Century Sea-Level Rise." *Science* 321 (5894): 1340–43.

Pidwirny, M. 2006. "Causes of Climate Change." In *Fundamentals of Physical Geography.* 2nd ed. http://www.physicalgeography.net/fundamentals/7y.html.

Schmidt, G. A. 2007. "The Physics of Climate Modeling." *Physics Today*, January, 72.

Schoof, C. 2010. "Ice-Sheet Acceleration Driven by Melt Supply Variability." *Nature* 468 (7325): 803–6. doi:10.1038/nature 09618.

Science Education Resource Center, Carleton College. 2008. "Paleoclimatology: How Can We Infer Past Climates?" Accessed December 1, 2010. http://serc.carleton.edu/microbelife/topics /proxies/paleoclimate.html.

Sheffield, J., and E. F. Wood. 2008. "Projected Changes in Drought Occurrence Under Future Global Warming from Multi-model, Multi-scenario, IPCC AR4 Simulations." *Climate Dynamics* 31 (1): 79–105.

Solomon, S., G.-K. Plattner, R. Knutti, and P. Friedlingstein. 2009. "Irreversible Climate Change due to Carbon Dioxide Emissions." *PNAS* 106 (6): 1704–9. doi:10.1073/pnas.0812721106.

Steffen, K., P. U. Clark, J. G. Cogley, D. Holland, S. Marshall, E. Rignot, and R. Thomas. 2008. "Rapid Changes in Glaciers and Ice Sheets and Their Impacts on Sea Level." In *Abrupt Climate Change*. Synthesis and Assessment Product 3.4, 60–142. Reston, Va.: U.S. Geological Survey.

Steig, E. J., D. P. Schneider, S. D. Rutherford, M. E. Mann, J. C. Comiso, and D. T. Shindell. 2009. "Warming of the Antarctic Ice-Sheet Surface Since the 1957 International Geophysical Year." *Nature*, January 22. doi:10.1038/nature07669.

Stroeve, J., M. M. Holland, W. Meier, T. Scambos, and M. Serreze. 2007. "Arctic Sea Ice Decline: Faster Than Forecast." *Geophysical Research Letters* 34, L09501. doi:10.1029/2007GL029703.

Stuiver, M., R. L. Burk, and P. D. Quay. 1984. "$^{13}C/^{12}C$ Ratios and the Transfer of Biospheric Carbon to the Atmosphere." *Journal of Geophysical Research* 89:11731–48.

Velicogna, I. 2009. "Increasing Rates of Ice Mass Loss from the Greenland and Antarctic Ice Sheets Revealed by GRACE." *Geophysical Research Letters* 36, L19503. doi:10.1029/2009 GL040222.

Way, R. 2010. "Is Antarctica Losing or Gaining Ice?" *Skeptical Science* (blog). http://www.skepticalscience.com/antarctica -gaining-ice-basic.htm.

Wigley, T. M. L. 1983. "The Pre-industrial Carbon Dioxide Level." *Climatic Change* 5 (4): 315–20. doi:10.1007/BF00140798.

Williams, S. J., B. T. Gutierrez, J. G. Titus, S. K. Gill, D. R. Cahoon, E. R. Thieler, K. E. Anderson, D. FitzGerald, B. Burkett, and J. Samenow. 2009. "Sea-Level Rise and Its Effects on the Coast." In U.S. Climate Change Science Program, *Coastal Elevations and Sensitivity to Sea-Level Rise: A Focus on the Mid-Atlantic Region.* Synthesis and Assessment Product 4.1. Washington, D.C.: U.S. Environmental Protection Agency.

Zachos, J., M. Pagani, L. Sloan, E. Thomas, and K. Billups. 2001. "Trends, Rhythms, and Aberrations in Global Climate 65 Ma to Present." *Science* 292 (5517): 686–93. doi:10.1126/science.1059412.

IMPACTS OF CLIMATE CHANGE

Environmental Protection Agency. *Climate Change Indicators in the United States.* Accessed December 1, 2010. http://www.epa.gov/climatechange/indicators.html.

U.S. Climate Change Science Program. 2009. *Thresholds of Climate Change in Ecosystems: A Report by the U.S. Climate Change Science Program and the Subcommittee on Global Change Research.* U.S. Geological Survey, Department of the Interior, Washington, D.C.

Extreme Weather and Tropical Storms

Bachelor, R. E. 2010. "Hurricane Demographics Index Destruction Level." May. http://www.suite101.com/content/hurricane-demographics-index-destruction-level-a242581.

Committee on Stabilization Targets for Atmospheric Greenhouse Gas Concentrations, National Research Council. 2011. *Climate Stabilization Targets: Emissions, Concentrations, and Impacts over Decades to Millennia.* Washington, D.C.: National Academies Press. http://www.nap.edu/openbook.php?record_id=12877.

Dai, A. 2011. "Drought Under Global Warming: A Review." *Wiley Interdisciplinary Reviews: Climate Change* 2 (1): 45–65. doi:10.1002/wcc.81.

Intergovernmental Panel on Climate Change. 2012. "Managing the Risks of Extreme Events and Disasters to Advance Climate Change Adaptation (SREX)." http://ipcc-wg2.gov/SREX/.

Knutson, T. R., J. L. McBride, J. Chan, K. Emanuel, G. Holland, C. Landsea, I. Held, J. P. Kossin, A. K. Srivastava, and M. Sugi. 2010. "Tropical Cyclones and Climate Change." *Nature Geosciences* 3 (3): 157–63.

Pielke, R. A., Jr., J. Gratz, C. W. Landsea, D. Collins, M. A. Saunders, and R. Musulin. 2008. "Normalized Hurricane Damage in the United States: 1900–2005." *Natural Hazards Review*, February.

U.S. Task and Rescue Service. "Hurricanes." Accessed December 1, 2010. http://www.ussartf.org/hurricanes.htm.

Sea-Level Rise

Australian Academy of Science. 2010. "Getting into Hot Water—Global Warming and Rising Sea Levels." *NOVA: Science in the News*. Accessed December 1, 2010. http://www.science.org.au/nova/082/082key.htm.

Radić, V., and R. Hock. 2011. "Regionally Differentiated Contribution of Mountain Glaciers and Ice Caps to Future Sea-Level Rise." *Nature Geoscience* 4 (2): 91–94. doi:10.1038/NGEO1052.

Vermeer, M., and S. Rahmstorf. 2009. "Global Sea Level Linked to Global Temperature." *PNAS* 106 (51): 21527–32. doi:10.1073pnas.0907765106.

Impacts on Plants and Animals

"Caribbean Coral Die-Off May Be Worst Ever." 2010. *Climate Progress*, October 20. http://climateprogress.org/2010/10/20/coral-bleaching-die-off-worst-ever/.

Ceballos, G., A. García, and P. R. Ehrlich. 2011. "The Sixth Extinction Crisis Loss of Animal Populations and Species." January 13. http://www.muracco.com/2011/01/the-sixth-extinction-crisis-loss-of-animal-populations-and-species/.

Church, J., and N. White. 2011. "Sea-Level Rise from the Late

19th to the Early 21st Century." *Surveys in Geophysics* 32 (4–5): 585–602. doi:10.1007/s10712-011-9119-1.

Coffman, K. 2011. "Bark Beetle Infestation Grows in Colorado, Wyoming." Reuters, January 23. http://www.reuters.com/article/2011/01/23/us-beetles-forests-idUSTRE70M28S20110123.

Division of Forestry, Alaska Department of Natural Resources. What's Bugging Alaska's Forests? Spruce Bark Beetle Facts and Figures. http://forestry.alaska.gov/insects/sprucebarkbeetle.htm.

Environmental Protection Agency. 2011. "Climate Change Effects on Coral." Accessed February 2, 2011. http://www.epa.gov/climatechange/effects/eco_coral.html.

Gardner, T. A., I. A. Côté, J. A. Gill, A. Grant, and A. R. Watkinson. 2003. "Long-Term Region-Wide Declines in Caribbean Corals." *Science* 301 (5635): 958–60. doi:10.1126/science.1086050.

Goldenberg, S. 2010. "A Home from Home: Saving Species from Climate Change." *Guardian*, February 12. http://www.guardian.co.uk/environment/2010/feb/12/saving-species-climate-change.

Heller, N. E., and E. S. Zavaleta. 2008. "Biodiversity Management in the Face of Climate Change: A Review of 22 Years of Recommendations." *Biological Conservation*, November 21.

Marinelli, J. 2010. "Guardian Angels." *Audubon*, May/June. http://www.audubonmagazine.org/features1005/activism.html.

Markey, S. 2006. "Global Warming Has Devastating Effect on Coral Reefs, Study Shows." *National Geographic News*, May 16. http://news.nationalgeographic.com/news/2006/05/warming-coral.html.

McLachlan, J. S., J. J. Hellmann, and M. W. Schwartz. 2007. "A Framework for Debate of Assisted Migration in an Era of Climate Change." *Conservation Biology* 21 (2): 297–302.

Menzel, A., and V. Dose. 2005. "Analysis of Long-Term Time Series of the Beginning of Flowering by Bayesian Function Estimation." *Meteorologische Zeitschrift* 14 (3): 429–34.

Parmesan, C. 2006. "Ecological and Evolutionary Responses to

Recent Climate Change." *Annual Review of Ecological Evolutionary Systems* 37:637–69.

———. 2007. "Influences of Species, Latitudes, and Methodologies on Estimates of Phenological Response to Global Warming." *Global Change Biology* 13 (9): 1860–72.

Parmesan, C., and G. Yohe. 2003. "A Globally Coherent Fingerprint of Climate Change Impacts Across Natural Systems." *Nature* 421 (6918): 37–42.

Ridgwell, A., and D. N. Schmidt. 2010. "Past Constraints on the Vulnerability of Marine Calcifiers to Massive Carbon Dioxide Release." *Nature Geoscience* 3 (3): 196–200. doi:10.1038/ngeo755.

Rozell, N. 2004. "Bark Beetles Take Connecticut-Size Bite out of Alaska." *Alaska Science Forum*, February 26. http://www2.gi.alaska.edu/ScienceForum/ASF16/1688.html.

Saino, N., D. Rubolini, E. Lehikoinen, L. V. Sokolov, A. Bonisoli-Alquati, R. Ambrosini, G. Boncoraglio, and A. P. Møller. 2009. "Climate Change Effects on Migration Phenology May Mismatch Brood Parasitic Cuckoos and Their Hosts." *Biology Letters* 5 (4): 539–41. doi:10.1098/rsbl.2009.0312.

Williams, J. W., and S. T. Jackson. 2007. "Novel Climates, No-Analog Communities, and Ecological Surprises." *Frontiers in Ecology and the Environment* 5 (9): 475–82. doi:10.1890/070037.

Impacts on Humans

Burke, M. B., E. Miguel, S. Satyanath, J. A. Dykema, and D. B. Lobell. 2009. "Warming Increases the Risk of Civil War in Africa." *PNAS* 106 (49): 20670–74. doi:10.1073/pnas.0907998106.

"Climate Refugee 'Crisis' Will Not Result in Mass Migration—New Research." 2011. *Ecologist*, February 4. http://www.theecologist.org/News/news_round_up/757943/climate_refugee_crisis_will_not_result_in_mass_migration_new_research.html.

De Souza, R.-M. 2004. "In Harm's Way: Hurricanes, Population Trends, and Environmental Change." Population Reference Bureau. http://www.prb.org/Articles/2004/InHarms

WayHurricanesPopulationTrendsandEnvironmentalChange.aspx.

McMichael, A. J., R. E. Woodruff, and S. Hales. 2006. "Climate Change and Human Health: Present and Future Risks." *Lancet* 367 (9513): 859–69. doi:10.1016/S0140-6736(06) 68079-3.

"One Billion Homeless by 2050, Says Christian Aid." 2007. *Ecologist*, May 14. http://www.theecologist.org/News/news_round_up/269141/1_billion_homeless_by_2050_says_christian_aid.html.

Raleigh, C., and H. Urdal. 2007. "Climate Change, Environmental Degradation, and Armed Conflict." *Political Geography* 26 (6): 674–94. doi:10.1016/j.polgeo.2007.06.005.

Roy, S. B., L. Chen, E. Girvetz, E. P. Maurer, W. B. Mills, and T. M. Grieb. 2010. *Evaluating Sustainability of Projected Water Demands Under Future Climate Change Scenarios.* Tetra Tech Inc. Report for Natural Resources Defense Council. http://rd.tetratech.com/climatechange/projects/nrdc_climate.asp.

World Hunger Education Service. 2011. 2011 World Hunger and Poverty Facts and Statistics. http://www.worldhunger.org/articles/Learn/world%20hunger%20facts%202002.htm.

SOLUTIONS AND CLIMATE-ENERGY CONNECTION

Assisted Migration (Assisted Colonization, Managed Relocation) and Rewilding of Plants and Animals in an Era of Global Warming. 2011. http://www.torreyaguardians.org/assisted-migration.html.

Dooley, J. J., C. L. Davidson, and R. T. Dahowski. 2009. *An Assessment of the Commercial Availability of Carbon Dioxide Capture and Storage Technologies as of June 2009.* U.S. Department of Energy. www.pnl.gov/main/publications/external/technical_reports/PNNL-18520.pdf.

Energy Information Administration. 2007. *Annual Energy Outlook 2007.* ftp://tonto.eia.doe.gov/forecasting/0383(2007).pdf.

Environmental Protection Agency. 2011. Renewable Fuel Standard. http://www.epa.gov/otaq/fuels/renewablefuels/index.htm.

Hill, J., E. Nelson, D. Tilman, S. Polasky, and D. Tiffany. 2006.

"Environmental, Economic, and Energetic Costs and Benefits of Biodiesel and Ethanol Biofuels." *PNAS* 103 (30): 11206–10. http://www.pnas.org/content/103/30/11206.full.pdf.

International Energy Agency. 2006. *World Energy Outlook 2006.* http://www.iea.org/weo/2006.asp.

Massachusetts Institute of Technology. 2007. *The Future of Coal.* http://web.mit.edu/coal/The_Future_of_Coal.pdf.

Portland Cement Association. *Technical Brief: Green in Practice 102—Concrete, Cement, and CO_2*. Accessed December 1, 2010. http://www.concretethinker.com/technicalbrief/Concrete-Cement-CO_2.aspx.

List of Outside Referees

In addition to review by Climate Central scientists, every chapter of this book has been reviewed by at least one referee, chosen for his or her expertise and professional reputation in a relevant area of climate or climate-related science. The referees are, in alphabetical order:

Cristina Archer, Associate Professor, College of Natural Resources, California State University, Chico

David Archer, Professor of Geophysical Sciences, University of Chicago

David Bahr, Research Affiliate, Institute of Arctic and Alpine Research, University of Colorado, Boulder

John Church, Fellow, Commonwealth Scientific and Industrial Research Organisation (Australia)

Keith Dixon, Research Scientist, Geophysical Fluid Dynamics Laboratory, National Oceanic and Atmospheric Administration

David R. Easterling, Chief of the Scientific Services Division, National Climatic Data Center, National Oceanic and Atmospheric Administration

Kristie Ebi, Consultant, Intergovernmental Panel on Climate Change

Kerry Emanuel, Professor of Meteorology, MIT

Chris Forest, Associate Professor of Climate Dynamics, Department of Meteorology, Pennsylvania State

Peter Gleckler, Research Scientist, Lawrence Livermore National Laboratory

Nancy Grumet Prouty, Research Geologist, U.S. Geological Survey

Matthew Huber, Professor, Climate Change Research Center, Purdue University

Jeff T. Kiehl, Climate Change Research Section Head and Senior Scientist, Climate and Global Dynamics Division, National Center for Atmospheric Research

Thomas R. Knutson, Research Meteorologist, Geophysical Fluid Dynamics Laboratory, National Oceanic and Atmospheric Administration

Josh Lawler, Associate Professor, School of Forest Resources, University of Washington

Scott Loarie, Department of Global Ecology, Carnegie Institution for Science

David Lobell, Assistant Professor, School of Earth Sciences, Stanford University

Camille Parmesan, Professor of Integrative Biology, University of Texas, Austin

Kelly Redmond, Deputy Director, Western Regional Climate Center

Richard Somerville, Distinguished Professor Emeritus and Research Professor, Scripps Institution of Oceanography, University of California, San Diego

Andrew Weaver, Professor, School of Earth and Ocean Sciences, University of Victoria, Canada

Christopher Yang, Assistant Project Scientist, Institute of Transportation Studies, University of California, Davis